AF251979

ADVANCES IN BIOPHYSICS, Volume 2, 1971

Vol. 2, 1971

ERRATA for Vol. 1

Page	Column	Line	For	Read
90	Fig. 18	legend ↑ 1	×, RNase N_1; $\triangle$, RNase U_1.	×, RNase U_1; $\triangle$, RNase N_1

ADVANCES IN BIOPHYSICS

Masao Kotani, Editor

Editorial Board
Setsuro Ebashi, Kazutomo Imahori, Toshizo Isemura,
Yasuji Katsuki, Jin-ichi Nagumo, Haruhiko Noda,
Fumio Oosawa, Ei Teramoto, Jun-ichi Tomizawa

Secretary
Koscak Maruyama
Biological Institute, University of Tokyo,
Meguro, Tokyo

UNIVERSITY OF TOKYO PRESS
Tokyo
UNIVERSITY PARK PRESS
Baltimore, London, and Tokyo
1971

Published jointly by
UNIVERSITY OF TOKYO PRESS
Tokyo
and
UNIVERSITY PARK PRESS
Baltimore, London, and Tokyo
ISBN 0-8391-0597-5
Library of Congress Catalogue Card No. 76-118663

PREFACE

Volume II is the second issue of "Advances in Biophysics," which is published, in principle, semiannually by the University of Tokyo Press in cooperation with University Park Press. The Biophysical Society of Japan is the scientific backer of the journal, and liaison with the Society is handled by the Editor and the Secretary.

As mentioned in the Preface to Volume I, "Advances in Biophysics " will include scientific papers in biophysics, and also those in related disciplines which are interesting and stimulating from the biophysical point-of-view. Each paper is intended to present systematically an overall account of an author's work on a specified subject in such a way as to be understandable to interested readers in a wide circle, who do not necessarily specialize in biophysics, without the necessity of their referring to other papers by the author and others. Pertinent background information and short reviews of related studies by other scientists will be included.

All five papers included in Volume II were solicited by the Editorial Board. It is not our intention, however, to exclude unsolicited contributions; on the contrary, the Editorial Board welcomes voluntary contributions suitable for publication by scientists in any country. Those wishings to contribute papers are requested to write in advance to the Editor or the Secretary.

We regret that the publication of Volume II was delayed due to unavoidable editorial reasons. Future volumes will appear semiannually as originally intended; accordingly, Volume III is expected to come out in fall this year.

Masao Kotani
Editor

CONTENTS

Preface v

PHASE TRANSITION IN MEMBRANE WITH REFERENCE
TO NERVE EXCITATION

Y. Kobatake, I. Tasaki and A. Watanabe 1

   I. Experimental Manipulation of the Milieu Surrounding the
     Squid Axon Membrane 2
  II. Asymmetry of the Axon Membrane..................... 4
 III. Electrochemical Properties of Axon Membrane at Rest.... 5
  IV. Abrupt Depolarization 9
   V. Axon Membrane in the Excited State 12
  VI. Stability of Resting and Active States of Axon Membrane.. 14
 VII. Effects of Temperature Changes on Excitability 15
VIII. Structure and Model for Axon Membrane 16
  IX. Two Stable States of Axon Membranes 20
   X. Optical Properties of Axon Membranes.................. 22
  XI. Conclusion 29

ONE-ELECTRON AND TWO-ELECTRON TRANSFER
MECHANISMS IN ENZYMIC OXIDATION REDUCTION
REACTIONS

I. Yamazaki 33

A Tentative Definition of One-Electron and Two-Electron Transfer
Mechanism at the Surface of Enzyme Molecules 36
Metal Enzymes (Oxidases and Peroxidases) 43
  1. Horseradish peroxidase 43
  2. Laccase and ascorbate oxidase 47

viii

   3.   Tyrosinase 49
Flavoproteins 50
   1.   Flavoproteins in electron transport systems 51
   2.   NAD(P)H dehydrogenase (DT diaphorase) 56
   3.   Xanthine oxidase and related flavoprotein oxidases 58
   4.   Flavoprotein oxidases 60
Dehydrogenases 61
Classification of the Electron Transfer Mechanism in the Enzymic
Oxidation-Reduction Reactions on the Basis of the Formation of
Free Radicals Derived from Donor and Acceptor 62
Discussion 66
Summary 72

THE ELECTROGENIC SODIUM PUMP   K. Koketsu 77

   1.   The concept of the electrogenic sodium pump 77
   2.   Membrane potential and electrogenic sodium pump 80
Experimental Evidence for an Electrogenic Sodium Pump 82
   1.   Resting potential 82
   2.   Action potential 95
   3.   Postsynaptic potential 104
Summary 108

STRUCTURE OF TROPOMYOSIN AND ITS CRYSTAL
                 T. Ooi and S. Fujime-Higashi 113

Isolation and Purification 114
Physicochemical Properties 116
   1.   Polymerization-depolymerization 116
   2.   Reversible equilibrium of tropomyosin polymers 121
   3.   Tropomyosin as a polyelectrolyte 123
   4.   Monomeric unit of tropomyosin 124
Biochemical Studies of Tropomyosin 127
Crystals of Tropomyosin 129
   1.   Crystallization 129
   2.   Electron microscopic observations 131
   3.   Polymorphism of the crystal 139
   4.   Polymorphic transition between various crystal forms ... 144

Interaction with Other Muscle Proteins 145
   1. Actin 145
   2. Troponin 145
   3. Other proteins 148
Possible Structure of Tropomyosin Molecule 149
Location of Tropomyosin in Myofibril 150
Summary 151

STOCHASTIC THEORY OF REACTION KINETICS

E. Teramoto, N. Shigesada, H. Nakajima and K. Sato   155

I. General Formula 156
   1. Introduction 156
   2. Definitions and mathematical formula 157
   3. Reaction rate function and master equation 159
   4. The specification of state 163
II. Bimolecular Reactions in a Liquid Medium 164
   5. Diffusion controlled reaction kinetics 164
   6. Mathematical formula 166
   7. Reduced formulas 168
   8. Multi-dimensional diffusion equation with pair absorbing
      interaction 171
   9. Cluster expansion 172
   10. Binary collision expansion 175
   11. Binary kernels in momentum representation 180
   12. Approximate results. 184
III. Computer Simulation 188
   13. Introduction 188
   14. The model 189
   15. Some typical results. 191
   16. Discussion 195
Summary 198

CONTRIBUTOR SKETCHES 201
PUBLISHED PAPERS 205
FORTHCOMING PAPERS 205

Advan. in Biophys., Vol. 2, pp. 1–31 (1971)

PHASE TRANSITION IN MEMBRANE WITH REFERENCE TO NERVE EXCITATION

YONOSUKE KOBATAKE,* ICHIJI TASAKI and AKIRA WATANABE

Laboratory of Neurobiology, National Institute of Mental Health, Bethesda, Maryland, U.S.A.

Elucidation of the processes of nerve excitation on a physicochemical basis is essential for an understanding of the functioning of the nervous system.   Because there are many undefinable quantities in a system consisting of an excitable membrane and its natural environment, however, there is frequently serious ambiguity in interpreting the results of physical measurements carried out on nervous tissues.   Due to this ambiguity and to the ease with which high time resolution measurements of the electrical concomitants of the excitation process can be made, prevalent theories (*1*) of excitability are concerned mainly with a mathematical description of membrane currents and potentials.   Such theories do not rely upon a well-defined physicochemical model of intramembrane mechanisms and are, therefore, not expected to yield any definitive information about the molecular processes involved in excitation.

The fluid media surrounding the axon membrane are, under ordinary experimental conditions, very complex.   The external medium

---

* Permanent address : Faculty of Pharmaceutical Sciences, Hokkaido University, Sapporo, Japan

(blood, Ringer or sea water) contains a large number of uni- and divalent cations and anions. The internal medium (protoplasm) contains organic polyelectrolytes in addition to inorganic salts. The ensuing difficulties were eliminated several years ago when the perfusion technique was introduced into physiological studies of squid giant axons (2, 3). It became possible to remove the protoplasm almost completely without affecting the excitability of the axon. The excitation of the axon could now be analyzed with simple, well-defined solutions on both sides of the membrane, and the necessary and sufficient conditions for the maintenance of excitability could be defined. One of the important conclusions derived from these studies is that the process of excitation is accompanied by a drastic change in the macromolecular conformation of the membrane, triggered by cooperative cation exchange at fixed negative sites in the membrane.

Recently, another powerful technique has been developed which promises to further increase our understanding of the mechanism of excitation. Changes were detected, with specifically designed optical and electrical equipment, in the optical properties of various invertebrate nerves during the nerve impulse (4, 5); changes in turbidity, birefringence, and extrinsic fluorescence were studied also. Changes in thermal properties (6) have also been detected during nerve excitation. These optical and thermal studies offer additional evidence that the conformation of the nerve membrane macromolecules changes during excitation.

I. EXPERIMENTAL MANIPULATION OF THE MILIEU SURROUNDING THE SQUID AXON MEMBRANE

The natural milieu of squid axons contains about 450 mM NaCl, about 60 mM of the salts of divalent cations, mainly Ca and Mg ions, and a relatively small amount of other ion species. Although the external milieu of the axon membrane could be readily altered, a physicochemical approach to nerve excitation was not possible until the composition of the internal milieu of the axon could be controlled. It may be said that the physicochemical investigations on the excitation mechanism began with the introduction in 1961 of the technique of intracellular perfusion (2).

It is well known that complete removal of divalent cations from the fluid external to an axon causes loss of excitability. However, complete elimination of univalent cations externally does not always lead to a complete loss of excitability. The first record in Fig. 1 illustrates the action potentials of an unperfused squid giant axon immersed in sea water. When the external solution was switched to a solution containing 100 mM $CaCl_2$ as the sole electrolyte and the axon was internally perfused with a 30 mequiv./1 Na-phosphate solution (pH 7.3), the action potential shown in the middle of the figure was observed. Although these fluid media are not " normal " for the axon, the axon membrane did not change irreversibly. When the external medium was switched back to natural sea water and the internal medium to a 400 mequiv./1 K-phosphate, the axon again produced action potentials with normal size and duration as shown in Fig. 1 (right).

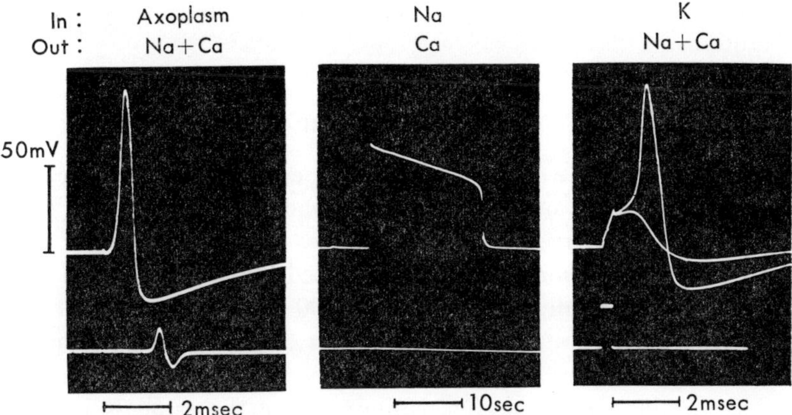

Fig. 1.   Oscillograph records of all-or-none action potentials in an axon under different conditions. *Left:* Record taken before initiation of internal perfusion ; the external medium contained 300 mM NaCl and 100 mM $CaCl_2$. *Middle:* Record taken approximately 12 min after the onset of internal perfusion with 10 mM sodium phosphate ; the external medium was 100 mM $CaCl_2$. *Right:* Record obtained from the same axon after switching the internal perfusion fluid to 400 mM K-phosphate and the external medium to the solution used in the left record ; sub- and suprathreshold responses are superposed. The time markers in the right and left records are 1 msec apart. The time marker in the middle record represents 10 sec. The stimulus duration was 0.1 msec for the left record, 100 msec for the middle record, and about 0.3 msec for the right record. Axon diameter: Approximately 400 μ, 21°C (Watanabe, Tasaki and Lerman, *Proc. Natl. Acad. Sci. U.S.*, **58**, 2246 (1967)).

In squid giant axons immersed in a 100 mM $CaCl_2$ solution, the ability to produce action potentials has been demonstrated with the salt of one of the following univalent cations internally: Na, Li, Cs, choline, tetramethylammonium, *etc*. (As we shall see later, either the phosphate or fluoride salts of these cations were most suitable for this type of experiment; sucrose or glycerol was added to the media to maintain the normal osmolarity.) Various salts of Sr or Ba can be used instead of $CaCl_2$ in the external medium (7). The action potentials obtained under these conditions are essentially similar to those shown in Fig. 1 (middle).

The analysis of the results obtained from a membrane system containing only two different species of cations is far easier and less amenable to misinterpretation than that from a system with three or more cation species. Therefore, a great emphasis will be placed in this article on the analysis of electrochemical properties observed under these " bi-ionic " conditions.

II. ASYMMETRY OF THE AXON MEMBRANE

When discussing the electrochemical properties of axon membranes, it is important to point out that the inner and outer surfaces of the axon membrane have very different properties. Surrounded with sea water externally, squid axons internally perfused with isotonic KCl solution, maintained excitability for only about 30 min. However, such axons perfused with K-glutamate, K-phosphate, or KF internally maintained excitability for many hours. Different survival times are obtained with different anion species perfused internally (at the same K-ion concentration). The survival time gave the following sequence of anion species (*8*):

$$F > \text{phosphate} > \text{glutamate, aspartate} > SO_4 > Cl > NO_3 > Br > I > SCN.$$

The fluoride ion is most favorable and the SCN ion least favorable for maintaining excitability. The same anion sequence was found with the salts of other cations. This sequence is identical to that of anions arranged according to their lyotropic numbers (*9*) and is consistent with the fact that polypeptides are the major components of the excitable membrane. Similarly, the order of favorability of internally perfused univalent cations for axons was determined by the same method as for anions, and was as follows:

$$Cs > Rb > K > NH_4 > Na > Li .$$

This type of sequence is the same as that found by Bungenberg de Jong for phosphate colloids (*10*).

In contrast with the difference among anions in the intracellular perfusion fluid, no significant differences in the effect on excitability were observed among different external anion species. It is well known that there is a marked difference in behavior among external cations. If one calls an external cation " favorable " when it tends to increase the action potential amplitude, the following cations have the sequence:

$$Li, Na > NH_4 > Cs > Rb > K .$$

(Note that the criterion for determining this cation sequence is different from that for internal cations and anions.)

As pointed out above, divalent cations are indispensable for maintaining the excitability of axon membranes. However, intracellular perfusion with a solution containing divalent cations leads to immediate and permanent loss of excitability. The effects of various univalent cations upon the action potential amplitude are different on the two sides of the axon membrane. All these experimental facts show that there is a marked difference between the physicochemical properties of the inner and outer layers of the axon membrane.

III.   ELECTROCHEMICAL PROPERTIES OF AXON MEMBRANES AT REST

Under continuous perfusion with an isotonic solution containing KF and sucrose, a squid giant axon maintains its excitability for many hours in an isotonic medium containing $CaCl_2$ and NaCl. When no electric current passes through the membrane, the transmembrane fluxes of all ion species in the system satisfy the following conditions:

$$J_{Na} + J_K + 2J_{Ca} - J_{Cl} - J_F = 0$$

where each flux J is expressed in $mole \cdot cm^{-2} \cdot sec^{-1}$.

Replacement of the chloride ion in the external medium with other anions, *e.g.*, Br, SO_4, ethylsulphate, *etc.*, does not change the functions of the axon. This fact implies that the axon membrane has a high fixed density of fixed negative charges. When the membrane has such a high density of fixed negative charges, the fluxes of anions should be small.

Furthermore, it has been found experimentally that the flux of the divalent cation, J_{Ca}, is far smaller than the fluxes of univalent cations (*11*). Consequently, the above equality is simplified approximately to

$$J_{Na} + J_K \cong 0.$$

Radio-tracer studies carried out under these experimental conditions reveal that the relation given above is approximately correct in the resting state of axon membranes. Namely, the resting efflux of the K-ion is roughly equal to the resting influx of the Na-ion, irrespective of the concentrations of K- and Na-ions. Under the conditions that the fluxes of anions and divalent cations are negligibly small, integration of the Nernst-Planck flux equations leads to (*12*):

$$J_{Na} = J_K = \frac{RTXu_K u_{Na} A}{(u_K - u_{Na})\delta} \ln (u_K/u_{Na}), \qquad (1)$$

where X is the fixed charge density of the membrane, δ and A are the effective thickness and area, of the membrane, and u_K and u_{Na} stand for the average mobilities of the K- and Na- ions in the membrane. (Note that A/δ is related to the compactness of the membrane.) In Eq. 1, R and T are the gas constant and the absolute temperature, respectively.

Under the same conditions, the electric resistance across the membrane is represented by the equation:

$$r = \frac{(u_K - u_{Na})\delta}{F^2 X u_K u_{Na} A} [\ln (u_K/n_{Na})]^{-1}, \qquad (2)$$

where F is the Faraday constant. A comparison between Eqs. 1 and 2 leads to the following equation:

$$r J_i = RT/F^2 \qquad (\text{for } i = K, Na) \qquad (3)$$

This equality is valid under the assumption that the membrane is highly permselective, *i.e.*, that the anion fluxes are negligibly small. The experimental data for the squid giant axon in the resting state show that Eq. 3 is valid, taking into account the possibility of experimental errors (*11*). This finding is consistent with the view that the membrane of the squid axon in the resting state is a cation exchanger membrane.

Now, let us consider a uniform cation exchanger membrane that separates two electrolyte solutions containing both univalent and divalent cations. The membrane is assumed to be permselective. The activities of uni- and divalent cations in the outer solution are a_1' and a_2', and those in the inner solution are a_1'' and a_2'', respectively. The integration of the Nernst-Planck equation yields the following equation for the membrane potential, $\Delta\varphi$ (13):

$$-\Delta\varphi = \frac{RT}{F}\left\{\ln\frac{a_1''}{a_1'}+\frac{u_2}{2u_2-u_1}\ln\frac{\gamma''+1}{\gamma'+1}+\frac{u_2-u_1}{2u_2-u_1}\ln\frac{u_2\gamma''-u_2+u_1}{u_2\gamma'-u_2+u_1}\right\}, \quad (4)$$

where γ' and γ'' denote the values of $(1+8KX\,a_2/a_1{}^2)^{1/2}$ at the membrane outer and inner surfaces, K is the equilibrium selectivity coefficient defined by

$$\bar{C}_2/\bar{C}_1{}^2 = Ka_2/a_1{}^2. \quad (5)$$

The other notations are the same as in the previous equations, except that the subscripts 1 and 2 refer to the univalent and divalent cations species, respectively. In Eq. 5, $\bar{C}_2$ and $\bar{C}_1$ are the average concentrations of divalent and univalent cations in the membrane phase which equilibrates with a solution containing divalent and univalent cations at the respective activities a_2 and a_1. In the derivation of Eq. 4, K, X and u_2/u_1 are assumed to be constant throughout the membrane. When the chemical composition of the solution on one side is held constant and the concentrations of uni- and divalent cations on the other side are varied, Eq. 4 is simplified to

$$\Delta\varphi = (RT/F)\left[\ln a_1'+\frac{u_2}{2u_2-u_1}\ln(\gamma'+1)+\frac{u_2-u_1}{2u_2-u_1}\ln(u_2\gamma'-u_2+u_1)\right]$$
$$+\text{const.} \quad (6)$$

The constant in Eq. 6 is related only to the composition of the solution of the side denoted by ''. It is easy to show that when either K or u_2/u_1 approaches infinity, Eq. 6 becomes

$$\Delta\varphi = (RT/2F)\ln a_2'+\text{const.} \quad (7)$$

In either case, the membrane potential is determined by the activity of the divalent cation species in the external solution only. Conversely, when K or u_2/u_1 approaches zero, Eq. 6 becomes

$$\Delta\varphi = (RT/F)\ln a_1'+\text{const.} \quad (8)$$

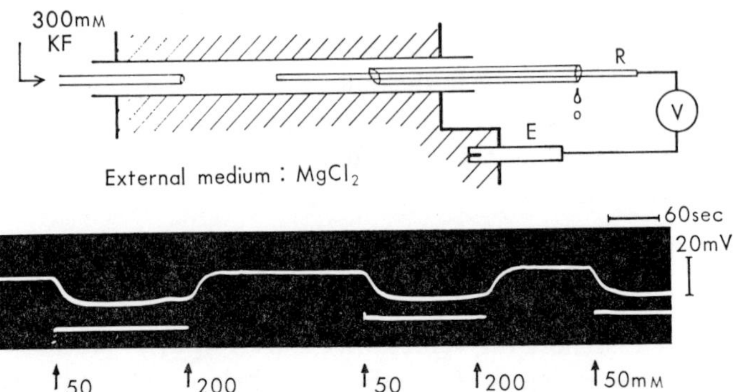

Fig. 2. *Top:* Schematic diagram showing the experimental setup used to determine the effect of the reduction of the external divalent cation concentration on the resting membrane potential ; E represents a calomel electrode and R a glass capillary electrode filled with 600 mM KCl. *Bottom:* Oscillograph records taken from an axon internally perfused with 300 mM KF (glycerol) solution (phosphate buffer). Downward deflections represent increased negativity within the axon. External media are prepared by mixing a 400 mM MgCl$_2$ (Tris-buffer) with 12% (v/v) glycerol (Tasaki, Watanabe and Lerman, *Am. J. Physiol.*, **213**, 1465 (1967)).

In this case, the activity of the univalent cation species in the external bulk solution determines the potential difference across the membrane. Therefore, observations of the dependence of the membrane potential on the concentration of cation species are expected to reveal whether or not the membrane has a very large selectivity coefficient K for divalent cation over univalent cation species.

The effect of the dilution of the external medium was examined by using internally perfused squid giant axons (*14*). An example of the results is shown in Fig. 2. A thoroughly cleaned axon was internally perfused with a solution containing 300 mM KF and 30 mM K-phosphate at pH=7.3. The external medium was a rapidly flowing solution containing 200 mM MgCl$_2$. The external medium was then switched to a rapidly flowing 50 mM MgCl$_2$ solution. As can be seen in Fig. 2, there was an immediate fall in the intracellular potential referred to the potential of the external medium. When the original 200 mM MgCl$_2$ solution was reintroduced, there was a prompt rise in the intracellular potential. In most cases, the effect of dilution was completely reversible. The magnitude of the potential change observed was 14 to 18 mV for a four-fold change in the concentration.   This observed value agrees very well with the value expected from Eq. 7.   Similar results were obtained when $CaCl_2$ was used instead of $MgCl_2$ in the external solution.   In the concentration range between 100 and 400 mM, the replacement of the Mg-ion with the Ca-ion produced no change in the membrane potential.

On the other hand, a change in concentration of the normally predominant Na-ion in the external medium at a fixed concentration of divalent cations hardly affects the intracellular potential in the resting state. For example, the addition of univalent cations, *e.g.*, Na-ions, to the external medium produces no significant change in the potential.   A tenfold variation of the Na-ion concentration, *i.e.*, from 10 to 100 mM in the presence of 200 mM $MgCl_2$, brings about only a few mV changes in the intracellular potential instead of the 58 mV expected from Eq. 8.   A similar result is observed for K-ions, provided that the K-ion concentration does not exceed a certain critical concentration, as will be discussed in detail in the subsequent section.

The experimental results mentioned above imply that the axon membrane in a resting state has a high selectivity for divalent cations over univalent cations.   (Note that the membrane potential is insensitive to the external univalent cation concentration when $K \gg 1$.)   In other words, the resting state of the axon membrane may be regarded as a divalent cation-rich state.   At lower concentrations of divalent cations, however, the membrane potential frequently becomes unstable;   the fluctuation of the membrane potential under these conditions is probably related to the process of action potential production, as will be discussed below.

IV.   ABRUPT DEPOLARIZATION

In squid giant axons internally perfused with Cs- (or Na-) phosphate and immersed in an external solution containing $CaCl_2$ as the sole electrolyte, the membrane potential can be abruptly changed when the K-ion added to the external medium exceeds a certain concentration. Figure 3 shows an example of the experiments of this type (*15*).   The axon is electrically excitable under these conditions as noted previously; but no electrical current was delivered to the axon during the following process.

The axon was kept in a steady state by rapid circulation of both internal and external solutions. At the onset, no univalent cation was present in the external medium. Then, portions of a nonelectrolyte (glycerol or sucrose) in the external medium were replaced with KCl and the external K-ion concentration was raised in small steps. No significant change in the membrane potential was observed when the external KCl concentration was raised successively from 0 to 5 mM. However, when the concentration of KCl reached a certain value, 10 mM in the case shown in Fig. 3, the intracellularly recorded potential increased

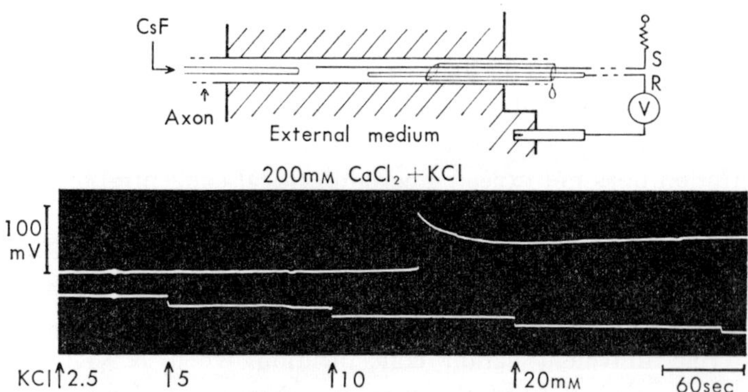

Fig. 3. Abrupt depolarization induced by external application of KCl. The KCl concentration in the rapidly flowing external fluid medium was raised by a factor of 2 at times marked by arrows. A silver wire electrode (S) was used to test the excitability of the axon before the experiment. R represents a glass pipette recording electrode. No electric stimuli were delivered while continuous recording of the membrane potential (V) was made. The potential jump produced by 10 mM KCl was 83 mV, and this was followed by a gradual potential fall of approximately 40 mV. Temperature, 16°C (Tasaki, Takenaka and Yamagishi, *Am. J. Physiol.*, **215**, 152 (1968)).

abruptly by more than 80 mV. This was followed by a gradual potential fall of about 40 mV. Once the potential jump had been observed, a further increase of the external KCl concentration caused no more sudden changes in the membrane potential. Instead, the potential changed smoothly with the KCl concentration and followed approximately Eq. 8: a tenfold variation of KCl concentration produced about a 58 mV change in the membrane potential.

In the field of electrophysiology, a rise in the internal potential level is called "depolarization." Following an "abrupt depolarization" caused by KCl, the axon cannot respond with an action potential to a stimulating (outward-directed) current through the membrane. When the axon membrane is depolarized, the lowering of the external KCl concentration produces "abrupt repolarization." The K-ion concentration required for repolarization is lower than that for depolarization.

When the external KCl concentration is only slightly above the critical concentration that caused an abrupt depolarization, an increase in the external Ca-ion concentration without changing the KCl concentration brings about repolarization.

As argued above, the observations indicate that the selectivity coefficient K of Eq. 5 suddenly changes from a very high value to a low value when the critical level of external KCl concentration is reached. In other words, the resting membrane is considered to be in a Ca-rich (or divalent cation-rich) state; and the depolarized membrane is regarded as being in a K-rich (or univalent cation-rich) state.

Measurements of the membrane impedance indicate that there is

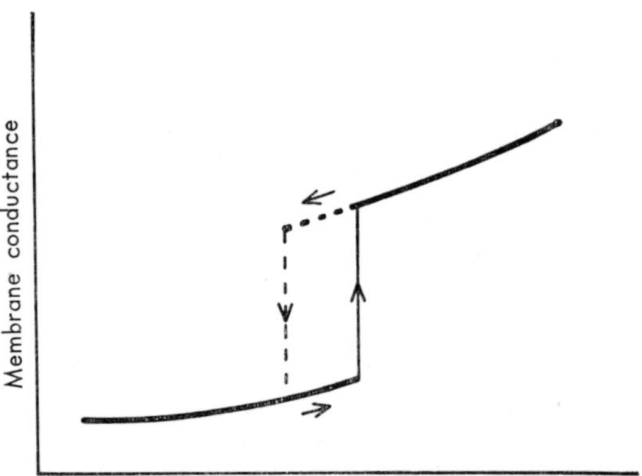

Fig. 4.   Simplified diagram showing abrupt changes in the membrane conductance produced by continuous changes in the mole fraction of the external univalent cation.   This diagram is based on the results from experiments with squid giant axons internally perfused with CsF or Na-phosphate.   Note the hysteresis loop in the diagram (Tasaki, Barry and Carnay, in "Physical Principles of Biological Membranes," ed. by F. Snell, J. Wolken, G. Iverson and J. Lam, Gordon and Breach, New York, (1970)).

a sudden and large fall in membrane resistance when the axon membrane is abruptly depolarized by the addition of K-ion to the external medium. This fall of the membrane resistance is followed by a gradual, partial restoration of the resistance with time, as will be shown later (see Fig. 7B). As noted in Sec. III, the electric resistance of a membrane is determined by the mobilities and concentrations of the mobile ions within the membrane. Hence the sudden fall in the membrane resistance associated with abrupt depolarization is a reflection of a sudden increase in the concentration of univalent cations in the membrane and/or a sudden decrease in the compactness of the membrane. The fact that a continuous rise in the external K-ion concentration produces an abrupt change in the membrane structure is a sign of the cooperativity of the process involved. In fact, a plot of the membrane resistance as a function of the mole fraction of the external univalent cations also shows an abrupt change, as shown in Fig. 4.

There are, in addition to the K-ion, other univalent cations which cause abrupt depolarization. The sequence of univalent cations that depolarize the membrane is

$$K > Rb > Cs > NH_4 > Na > Li .$$

The K-ion has the strongest depolarization power, *i.e.*, the lowest critical concentration for abrupt depolarization. A difference in the chemical species of anions in the external medium does not appreciably alter the observed depolarization power of univalent cations. It is interesting that the above ion sequence is the reverse of that for the favorability of external cation species, and that the sequence is similar to that determined for the so-called sodium-pump mechanism (*16*). The reason for this similarity will be explained in the later section.

V. AXON MEMBRANE IN THE EXCITED STATE

The excitability of a squid giant axon can be maintained for many hours with an external medium containing 200 to 500 mM NaCl and 50 to 100 mM $CaCl_2$ and an internal perfusion fluid consisting of 400 mM KF. Then the amplitudes of the action potential vary with the external Na-ion concentration in accordance with Eq. 8. Note, however, that such an amplitude-augmenting effect is not limited to the Na-ion. It is known that the excitability of the axon can be maintained when the Na-ions in the external medium are replaced with polyatomic cations, such as hydrazinium, guanidinium, NH_4, or other appropriate univalent cations. The amplitude of the action potential varies logarithmically with the concentration of these polyatomic cations.

Changing anions in the external solution does not affect the excitability or the amplitudes of the action potential. In other words, the axon membrane behaves like a reversible, univalent cation-sensitive electrode in the excited state, irrespective of the external cation species. In this sense, the excited state of the axon membrane is considered to be equivalent to the depolarized state of the membrane discussed in the previous section.

Isotope measurements have been carried out on axons internally perfused with a K-salt solution and immersed in a solution containing both NaCl and $CaCl_2$. Such measurements indicate that the fluxes of anions, J_{Cl}, and of divalent cations, $e.g.$, J_{Ca}, during the excited state are almost negligible compared with those of interdiffusing univalent cations, J_K and J_{Na}. Both J_K and J_{Na} increase by a factor of about 200 at the peak of an action potential compared with the levels in the resting state. The influx of Na-ions, J_{Na}, is roughly equal and opposite to the out-flux of K-ions, J_K.

Impedance measurements as well as determination of the current-voltage relationship of the axon in the resting and excited states show that the electric resistance of the axon decreases by a factor of about 200 at the peak of action potential. This implies that the electric resistance-flux product, $i.e.$, $r \times J_i$ ($i=$K or Na), remains roughly unaltered during excitation. It then follows from Eq. 3 that the axon membrane behaves as a permselective cation exchanger membrane in the excited state, as well as in the resting state. Thus, we may conclude that anions in the external medium are nearly perfectly excluded from the membrane phase both in the resting and excited states.

As noted previously, a stimulating current is directed outward through the axon membrane. Such a current tends to transport internal univalent cations, K-ions, into the axon membrane and to increase the mole fraction of the K-ions in the membrane. This increase in the mole fraction of K-ions is similar to that obtained with the addition of K-ions to the external medium. The favorable internal cations (K, Rb, Cs) have a stronger tendency to depolarize the membrane than the usual external univalent cations (Na, Li, *etc.*). Consequently, the transport of the internal, more depolarizing cations into the critical layer of the axon membrane by a stimulating current pulse is very effective in bringing the membrane to the depolarized state. Thus, it is considered that the initiation of an action potential by a stimulating current is an abrupt depolarization triggered by electrophoresis of the internal univalent cation into the membrane.

VI. STABILITY OF RESTING AND ACTIVE STATES OF AXON MEMBRANE

Both the resting and the excited states of an axon membrane are stable. This stability is examined by applying a small perturbation from an electric field across the membrane. Figure 5 illustrates the procedure

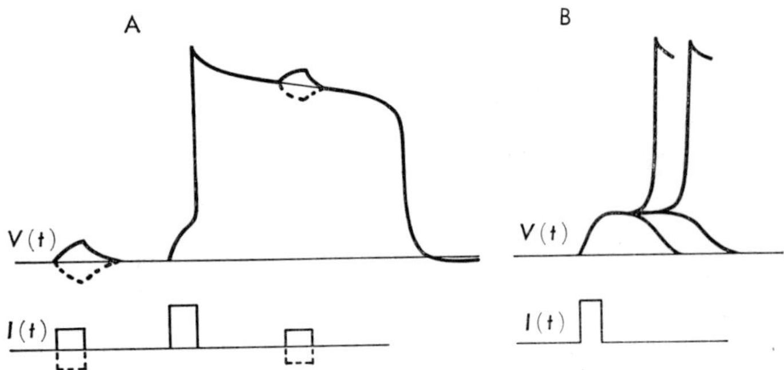

Fig. 5. A: Diagram showing the " stability " of the squid axon membrane in the resting state and during the plateau of a prolonged action potential. The lower trace, marked $I(t)$, shows the time courses of the rectangular current pulses applied to the axon. The upper trace, marked $V(t)$, shows the variation of the membrane potential produced by the current pulses. B: Diagram showing the variability in the potential-time curve for stimuli at threshold intensity; the four traces represent the membrane potentials observed at the same stimulus intensity (Tasaki, in " Nerve Excitation," Charles C. Thomas, Springfield, Ill., p. 106 (1968)).

for such an examination. The lower trace in the figure represents the time course of the rectangular current pulses applied to the axon membrane. The upper trace represents the resulting variations of the intracellularly recorded membrane potential. When the intensity of a perturbing current pulse is low, the transmembrane potential of the axon in the resting state varies exponentially with time. The potential variation brought about by an inward-directed current is approximately equal in magnitude (and the opposite in sign) to the variation caused by an outward-directed current pulse of the same intensity. The termination of the applied current leads to the return of the potential to the initial level. A similar potential change can be demonstrated during the plateau phase of a prolonged action potential (*17*).

When the current pulse applied to the axon membrane at rest raises the intracellular potential level by about 20 mV, an action potential is produced. After a current pulse of threshold intensity, the transmembrane potential is constant for a variable period of time before it approaches either the upper or the lower value of the membrane potential. In Fig. 5B, the four oscillograph traces show the time course of the membrane potentials with the same stimulus intensity. These observations indicate that both the upper (excited) and lower (resting) levels in the membrane potential are stable. Furthermore, the critical potential may be considered to be an intermediate, unstable level between the two stable levels. It is known that a pulse of inward-directed current delivered during the plateau of a prolonged action potential can produce a transition from the upper potential level to the lower. In order to produce this transition, the membrane potential has to be displaced beyond the unstable level between the upper and lower stable potential levels.

We shall now discuss the stability of the axon membrane that has been depolarized by application of KCl externally. When weak perturbing current pulses are applied to such a membrane, it is found that the depolarized state of the membrane is stable. Also, a strong pulse of outward-directed current does not evoke an action potential. But, a strong pulse of inward-directed current shifts the membrane potential from the depolarized level to the repolarized. Electric responses (*i.e.*, large, abrupt changes in the membrane potential) observed in depolarized axons are reversed in sign (*18*).

VII.   EFFECTS OF TEMPERATURE CHANGES ON EXCITABILITY

It has long been known that an action potential can be evoked by sudden cooling in various excitable tissues. This implies that the process in the axon membrane underlying depolarization (excitation) is exothermic, and that repolarization is endothermic. Since an endothermic reaction is encouraged by heating, a rise in temperature should lead to repolarization of the excited membrane. This was tested for a prolonged action potential by applying a brief heat pulse (*19*). It was found that the action potential could be terminated by a heat pulse in an all-or-none manner. This observation supports the view that the termination of the action potential (and, hence, repolarization) is a cooperative and endothermic reaction. Recent studies on temperature change during action potential also support this view (*6*).

VIII. STRUCTURE AND MODEL FOR AXON MEMBRANE

Since the adherent Schwann cell and connective tissue cannot be removed from the external surface of the 100 Å thick axonal membrane, it is difficult to chemically analyze the squid axon membrane. However, considering the experimental results described in the preceding sections, we may visualize a physical picture of the excitable membrane (*20*).

The excitable membrane is represented by a network of macromolecules held together, at least partly, by intra- and intermolecular salt-linkages between adjacent charged groups. The membrane may be subdivided into two distinct layers: The outer layer is compact with many fixed negatively charged groups on its component macromolecules; these negatively charged groups are probably carboxylates and confer cation exchanger properties on the outer layer. The inner layer of the membrane is less compact and contains a small number of fixed anionic sites, probably phosphates. The major diffusion barrier is considered to be the outer layer of the membrane.

In the outer layer, the complex formation between the fixed charged groups and the divalent cations from the external medium, as well as hydrogen and hydrophobic bonds, keeps the structure compact in the resting state. However, this structure is labile and can readily be disrupted by univalent cations, especially by the invasion into the outer layer of the univalent cations that have a strong depolarizing power, like the K-ion. The disruption leads to a decrease in compactness and to an increase in hydrophilicity of the outer membrane layer. The change in the membrane during excitation (or depolarization) may be attributed to this disruption. The decrease in compactness of the membrane in the depolarized state partly stems from the strong repulsive forces between the charged groups of the membrane skeleton released by disruption of complexes.   The rise in hydrophilicity may be due to the rearrangement of water molecules around the negatively charged sites exposed by the removal of divalent cations.

It is reasonable to assume that the structure of the membrane is not perfectly uniform along the axon surface.   The distribution of the depolarized and repolarized portion in the membrane is considered to fluctuate with time due to the thermal motion of ions and membrane macromolecules.   When the axon (as a whole) is in the resting state, the major portion of the membrane surface is compact and rich in divalent cations.   When the axon is excited, the major portion of the membrane surface is in a loose, univalent cation-rich state.

The potential across a membrane depends on both the salt compositions of the external and internal media and on the structure of the membrane.   Therefore, the membrane potential in the excited conformational state of the membrane differs from that in the resting state even when the chemical composition in the milieu stays constant.   Then, if the membrane is composed of a mixture of patches in the excited state and of patches in the resting state, the local membrane potential will vary from one position to another along the membrane surface.

A depolarized (excited) patch has a smaller negative membrane potential than a resting patch has.   This non-uniform distribution of patches with different emfs along the membrane surface creates local or eddy currents even when there is no net electric current across the membrane.   The local currents are inwardly directed in excited patches, and outwardly directed across resting patches.   The inward current through the excited patches of the membrane tends to induce repolarization by bringing Ca-ions into the membrane from the outer solution, while the outward current through the resting portion tends to bring about depolarization by carrying K-ions into the membrane from the axon interior.   As a consequence of this long range interaction between resting and excited portions of the membrane, an excitable axon which is as a whole in its resting state, is considered to possess active (excited) spots which appear and disappear spontaneously.

When the major cation species in the external medium are Na- and Ca-ions, the axon membrane can be in the resting state even at a relatively high Na-ion concentration because of the weak depolarizing power of the Na-ions. If K-ions invade the external surface of the membrane by the application of a stimulating current or by an increase of the KCl concentration in the outer solution, the membrane fraction in the depolarized state rapidly increases because of the strong depolarizing power of K-ions. When the fraction of the membrane in the excited state reaches a certain critical value, the membrane becomes predominantly active. A simple calculation based on the difference in membrane potential and conductance between the excited and resting states of the axon membrane indicates that the critical active fraction is less than 1% of the total membrane for the case of the squid axon membrane. This indicates that excitation of only 1% of the entire surface of the membrane is sufficient to throw the remaining 99% of the membrane into the excited state.

In the above discussion, it is assumed that every surface element (subunit) of the membrane is either in the excited (active) or the resting state (20–22). We now consider possible physicochemical bases for this property of the membrane macromolecules.

Barrer and Falconer (23) have shown that the X-ray diffraction pattern of Na-rich zeolite is different from that of K-rich zeolite. As the mole fraction of K-ion in a solution bathing Na-rich zeolite is continuously increased, the X-ray pattern changes. In the reverse direction where the K-rich form is converted to the Na-rich form, the transition occurs at a different mole fraction. A similar discontinuous process is encountered in the exchange between divalent and univalent cations in certain synthetic zeolites (24). These discontinuous ion-exchange processes can be explained in terms of statistical mechanics. When the occupancy of the neighboring sites by cations of the same kind is energetically more favorable than the occupancy of the adjacent sites by different cation species, a cooperative ion-exchange transition is expected. Making use of the Bragg-Williams approximation on the two dimensional lattice, we can derive an ion-exchange isotherm for a two cation species system (23).

An example of a theoretical ion-exchange isotherm for divalent-univalent cations on a two-dimensional lattice is illustrated in Fig. 6 (20). The abscissa represents the equivalent fraction of univalent cations in the solution phase, and the ordinate represents the equivalent fraction of univalent cations within the membrane phase. On the line marked ABC, the fraction of divalent cations in the membrane phase is much

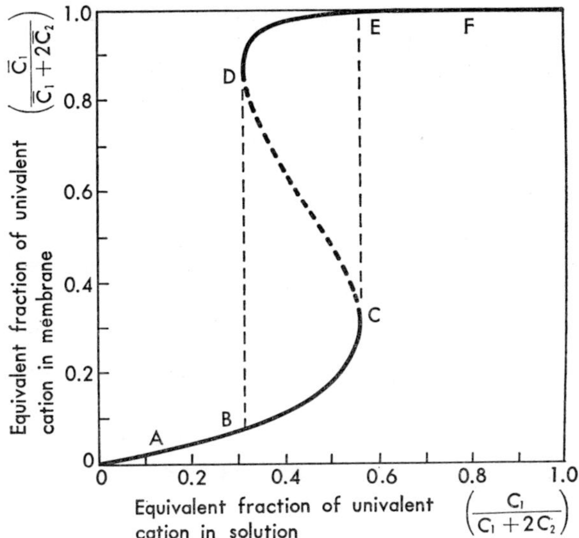

Fig. 6.   Theoretical ion exchange isotherm calculated for a cation exchanger membrane immersed in a salt solution containing univalent and divalent cations.   In the calculation, it was assumed that occupancy of two neighboring charged sites in the membrane by two cations of different valences is energetically unfavorable. C_1 and C_2 represent the concentrations in the solution of univalent and divalent cations, respectively; $\bar{C}_1$ and $\bar{C}_2$ represent the concentrations within the membrane.   When the equivalent fraction of the univalent cation in the solution is increased continuously from 0 to 1, the corresponding fraction in the membrane increases along the course 0, A, B, C, E, and F. When the equivalent fraction in the solution is decreased from 1 to 0, the corresponding fraction in the membrane changes along F, E, D, B, A, and 0 (Lerman, Watanabe and Tasaki, in "Neurosciences Research," ed. by Ehrenpreis and Solnitzky, Academic Press, New York, Vol. 2, p. 71 (1969)).

larger than that in the bathing solution phase, that is, the selectivity coefficient K defined by Eq. 5 is much larger than unity, $K \gg 1$.   When the external univalent cation fraction reaches C in Fig. 6, there is a discontinuous increase in the fraction of the univalent cations in the membrane phase.   A further increase in the fraction of the external univalent cation leads to a continuous increase in the intramembrane univalent cation concentration.   At this stage, the selectivity constant K is much smaller than unity, $K \ll 1$.

The isotherm shown in Fig. 6 implies that a hysteresis loop exists in this type of cation-exchange process. When the external univalent concentration is lowered continuously from the point marked F, a sudden fall in the intramembrane univalent fraction is expected to occur along the line D-B. The external univalent fraction at point B is much lower than that at point C. A hysteresis is also observed in the abrupt depolarization experiments on the squid giant axon membranes as discussed above (see Fig. 4).

IX. TWO STABLE STATES OF AXON MEMBRANES

The resting state of an axon membrane corresponds to the segment ABC in Fig. 6, where the negative sites of the axon membrane are predominantly occupied by divalent cations derived from the external solution. The depolarized or excited state of the membrane corresponds to the segment DEF. Here, the equivalent fraction of univalent cations in the membrane is greater than that of divalent cations. The discontinuous relationship between the external univalent cation concentration and the membrane potential in abrupt depolarization experiments, as well as in production of action potentials under bi-ionic situations, can thus be interpreted as representing a phase transition within the membrane associated with a cooperative ion-exchange process involving univalent and divalent cations at anionic membrane sites.

As mentioned in the previous section, the value of the selectivity coefficient K defined by Eq. 5 varies discontinuously from a very large value on the segment ABC to a small value on the segment DEF at point C. A sudden decrease in K, and thus a sudden increase in the univalent cation fraction at the outer membrane surface, causes a sudden rise in the intracellular potential, as seen in Eq. 4. Since the intramembrane mobility of univalent cations is greater than that of divalent cations, the sudden rise in the intramembrane univalent cation fraction is accompanied by an abrupt fall in membrane resistance.

When membrane resistance suddenly falls following transition, there is a corresponding rise in the cation fluxes as described in Eq. 3. This rise in the cation fluxes gradually decreases the concentration difference (across the membrane) of the cations contributing to the membrane potential, resulting in a gradual shift of the potential.

The results of impedance measurements on axons under bi-ionic conditions (shown in Fig. 7) indicate that membrane resistance progres-

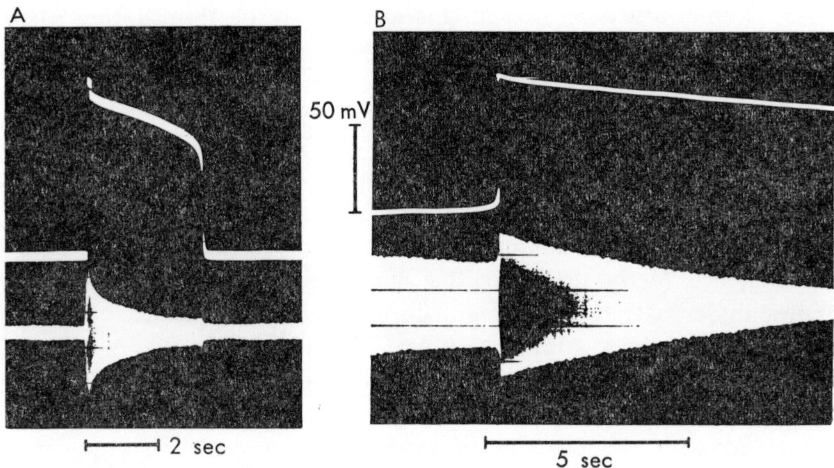

Fig. 7.   A :   Changes in the membrane impedance associated with the electrically induced action potential of an axon internally perfused with 25 mM CsF solution and immersed in 200 mM CaCl$_2$ solution.   The impedance bridge was balanced with the membrane impedance in the resting state.   The room temperature was 18°C.   B :   Changes in the membrane impedance produced by an alteration of the composition of the external medium from a mixture of 200 mM CaCl$_2$ and 25 mM CsCl to a mixture of 100 mM CaCl$_2$ and 450 mM CsCl (abrupt depolarization).   The temperature was 19°C.   Note that the sudden rises in the membrane potential were associated with simultaneous changes in the membrane impedance (Tasaki, Takenaka and Yamagishi, *Am. J. Physiol.*, **215**, 152 (1968)).

sively increases during action potential.   Thus the interdiffusion fluxes of cations through the membrane would gradually fall while the axon is in the excited state.   This fall of cation fluxes in turn raises the divalent cation fraction in the outer membrane surface.   Eventually, the membrane undergoes a transition from the univalent cation-rich state to the divalent cation-rich state (D-B transition in Fig. 6), bringing about an abrupt repolarization.   Thus the phenomena of abrupt depolarization and repolarization and, hence, the production of action potential are adequately explained on the basis of cooperative ion-exchange processes.

As was shown in Sec. VI, both the excited and resting states are stable in the sense that small perturbations do not trigger large changes in the properties of the membrane and that the membrane returns to the original state when the perturbations are removed.   For this reason, the logical framework described in this article may be termed the " two stable states theory."

X. OPTICAL PROPERTIES OF AXON MEMBRANES

Until quite recently, rapid physicochemical changes occurring during excitation in and near the axon membrane could be detected only by electrophysiological techniques. Now it is known that light scattering, birefringence (4, 5), and extrinsic fluorescence (5, 25) of axon membranes undergo transient changes during the action potential. These optical observations strongly support the view that the process of excitation is accompanied by a conformational change of the macromolecules within the axon membrane. In this article we are mainly concerned with the changes in extrinsic fluorescence, because such changes are readily interpreted on a molecular basis.

Under natural conditions, nerves from squid or spider crabs, do not show any detectable fluorescence in the visible range of the spectrum. When nerves are vitally stained with the appropriate fluorescent dyes and are irradiated with quasi-monochromatic light, the emission of fluorescent light by the nerve can be detected. When the stained nerves are stimulated electrically, a small transient change in the intensity of the fluorescent light occurs when the nerve impulse arrives at the irradiated portion of the nerve.

The fluorescent compounds listed in Table I give optical signals (25). Some of these dyes give rise to negative signals, representing a transient decrease in fluorescence during the excitation (negative sign in Table I). The names and chemical structures of dye molecules and the wave lengths of the excited quasi-monochromatic light used are listed in the first and second columns of the table. The signs and the relative magnitudes of the observed optical signals, which are between 5×10^{-4} and 10^{-5} times the fluorescence of the unstimulated nerve preparation, are given in the table for three different nerve preparations. The dyes were applied intracellularly in the case of squid giant axons; with squid fin nerves and crabn erves, staining was performed by immersion.

Nerves stained externally with Pyronin B and Rhodamine B gave rise to a transient decrease in fluorescence intensity during action potentials. It was not expected that intracellularly applied Pyronin B

TABLE I

Compounds Used to Demonstrate Fluorescence Changes in Nerves

Compound	Excitation wavelength in nm	Giant axon	Fin nerve	Crab nerve
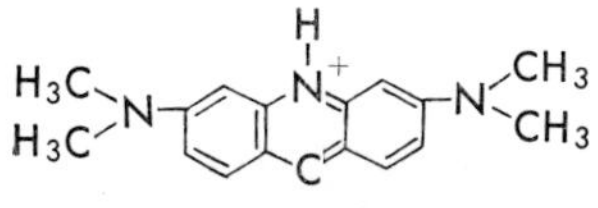 Acridine orange	465 (20)	+++	+++	+++
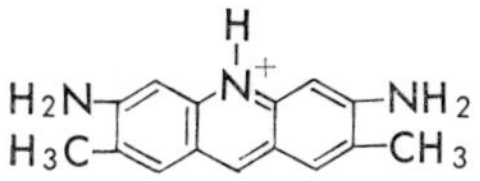 Acridine yellow	450 (20)	0	+	++
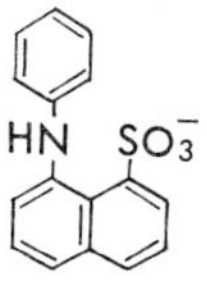 ANS	365 (10)	—	++	+++
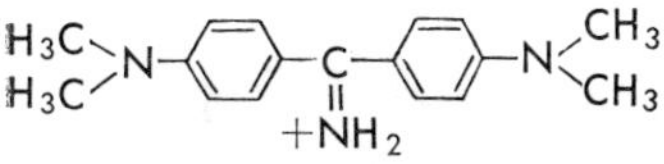 Auramine O	450 (5)	0	0	++
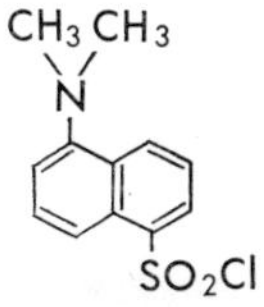 DNS	365 (10)	0	0	+
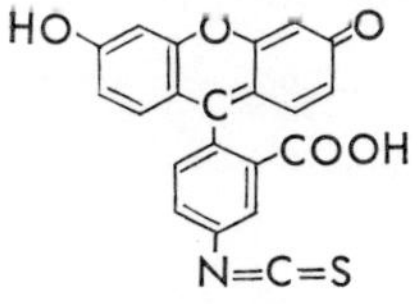 FIT	500 (20)	+ (ext)	++	++

continued

TABLE I

Compounds Used to Demonstrate Fluorescence Changes in Nerves (continued)

Compound	Excitation wavelength in nm	Giant axon	Fin nerve	Crab nerve
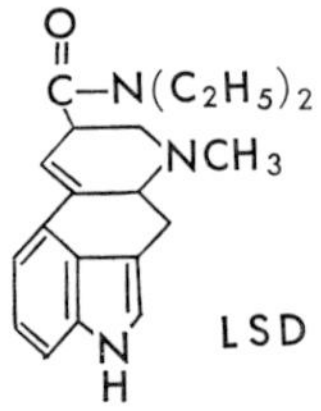 LSD	365 (10)	0	+	++
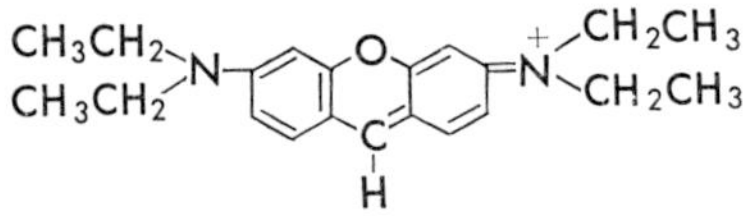 Pyronin B	550 (5)	++	—	—
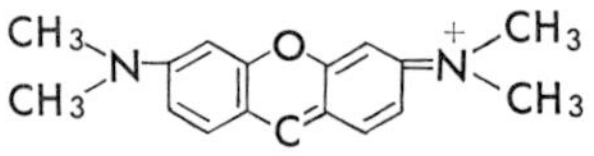 Pyronin Y	550 (5)	0	—	—
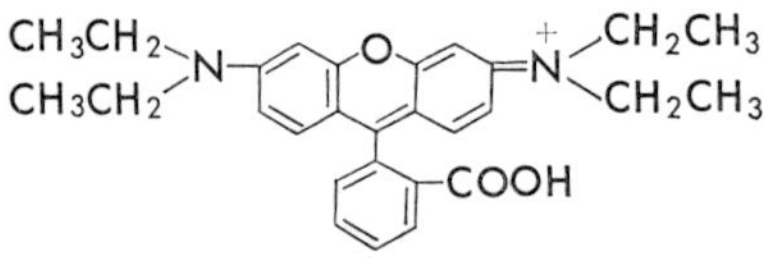 Rhodamine B	550 (5)	—	—	—
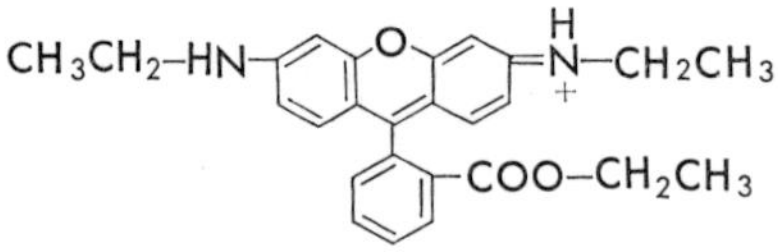 Rhodamine G	500 (20)	0		++

continued

TABLE I

Compounds Used to Demonstrate Fluorescence Changes in Nerves (continued)

Compound	Excitation wavelength in nm	Giant axon	Fin nerve	Crab nerve
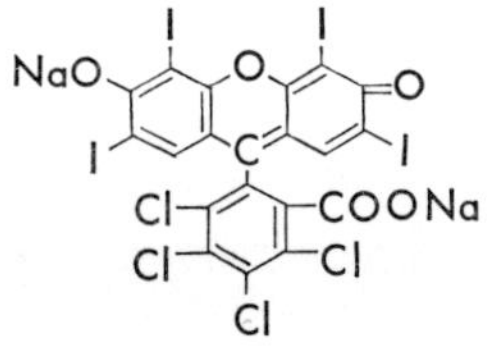 Rose bengal	550 (5)	0		++
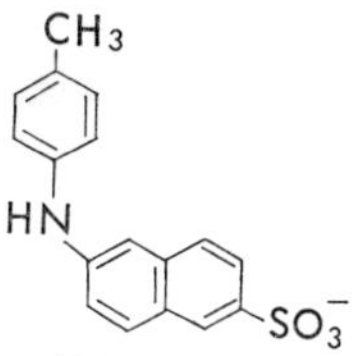 TNS	365 (20)	−		−

Symbols: ANS, 8-anilinonaphthalene-1-sulfonate; DNS, 1-dimethyl-aminonaph-thalene-5-sulfonyl chloride; FIT, fluorescein isothiocyanate; LSD, lysergic acid diethylamide; TNS, 2-p-toluidinylnaphthalene-sulfonate. The half band widths of filters used are given in parentheses. Strong, medium, and weak optical signals are shown by the difference in the number of + (increase) and − (decrease) symbols; the absence of a measurable signal is shown by 0 (adopted from Tasaki, Carnay and Wata-nabe, *Proc. Natl. Acad. Sci.*, **64**, 1362 (1969)).

would give rise predominantly to an increase in fluorescent light during excitation. It was also of great interest that the sign of the signal from squid giant axons perfused with ANS was opposite to that of the ANS signals obtained from squid fin and crab nerves.

The measurement of "*fluorescence polarization*" offers direct infor-mation about the viscosity of the medium surrounding fluorescent molecules. Figure 8 shows the principle of the technique used for measuring the fluorescence polarization. The white light from a strong incandescent lamp was converted into quasi-monochromatic light by an interference filter F_1. The monochromatic light was polarized by in-serting a polarizer P between F_1 and the nerve, so that the electric vector

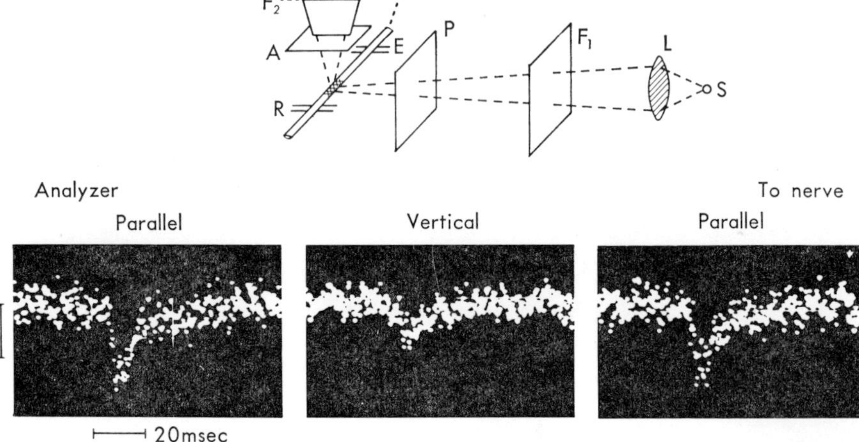

Fig. 8. *Top;* Schematic diagram showing the experimental setup used to measure changes in fluorescence polarization during excitation in a crab nerve stained with Pyronin B. The letter S represents the light source; L, lens; F_1, primary filter; F_2, secondary filter; M, photomultiplier tube; P, polarizer; A, analyzer; N, nerve; E, stimulating electrodes; R, recording electrodes. *Bottom;* Optical signals obtained with the principal axis of the analyzer (E-vector of fluorescent light) parallel to the long axis of the nerve (right and left records) and perpendicular to the nerve (middle record). The quasi-monochromatic light used for excitation of dye molecules was polarized with its E-vector in the direction parallel to the nerve. A CAT computer was used for recording. For right and left records, the vertical bar represents 2×10^{-4} times the background intensity. The middle record was taken under the same experimental conditions except that the analyzer was rotated by 90°. The temperature was 19°C (adopted from Tasaki, Carnay and Watanabe, *Proc. Natl. Acad. Sci. U. S.*, **64**, 1362 (1969)).

of the incident light wave was parallel to the longitudinal axis of the nerve. The light was focused on the stained nerve by a lens L. The fluorescent light from the nerve was detected at 90° to the incident beam with a photomultiplier tube; a secondary filter F_2 and an analyzer A were placed between the nerve and the photomultiplier. The analyzer could be rotated 90° so that fluorescent light, polarized either parallel or perpendicular, to the nerve could be detected. To improve the signal-to-noise ratio in recording, a digital computer was used for averaging multiple signals. The nerve was stimulated with brief pulses of electric current applied to the nerve near one end, and the action potentials were recorded externally at the other end.

Figure 8 (bottom) shows recordings from a crab nerve stained externally with Pyronin B. The dye molecules in the stained nerve were excited with a monochromatic polarized light 550 mμ in wave length.

Changes in the fluorescent light from the nerve were first measured with the analyzer parallel to the nerve axis (right and left records in Fig. 8). Thenthe analyzer was rotated 90° and the change in emitted light during the action potential was again recorded (middle record). It was found that the change in the fluorescent light with its electric vector parallel to the nerve is approximately twice as large as that polarized vertically to the nerve.

The degree of polarization of the fluorescent light from the resting nerve is defined by

$$P_r = \frac{I_r' - I_r''}{I_r' + I_r''},\tag{9}$$

where I_r' and I_r'' are the intensities of the fluorescent light polarized longitudinally and vertically relative to the nerve axis. The degree of polarization during excitation, P_a, is defined similarly in terms of the intensities, I_a' and I_a'', observed during an action potential:

$$P_a = (I_a' - I_a'')/(I_a' + I_a'').$$

It is difficult to measure P_a directly, because the simultaneous determination of I_a' and I_a'' during an action potential is required. However, the variation in polarization associated with nerve excitation, *i.e.*,

$$(P_a - P_r),$$

can be calculated by the use of the following expression:

$$P_a - P_r = \frac{1}{2}\left\{\frac{I_a' - I_r'}{I_r'} - \frac{I_a''}{I_r''}\frac{I_r''}{I_r''}\right\}(1 - P_r^2),\tag{10}$$

where subscripts a and r stand for the active (excited) and resting states of the nerve. The quantity $(I_a' - I_r')/I_r'$ in this equation represents the size of the optical signal observed with the principal axis of the analyzer parallel to the nerve; $(I_a'' - I_r'')/I_r''$ denotes the signal size obtained after the rotation of the analyzer through 90°. In the derivation of Eq. 10, the fact that the variation of the fluorescent light intensity during excitation is far smaller than unity, *i.e.*,

$$(I_\mathrm{a}-I_\mathrm{r})/I_\mathrm{r} \ll 1 ,$$

has been used.

In crab nerves stained with Pyronin B, the degree of polarization in the resting state, P_r, was found to be approximately 0.14. During excitation the intensity of the emitted light decreases; I' decreases much more than does I''. This means that the degree of polarization decreases.

In the example shown in Fig. 8, $(I_\mathrm{a}'-I_\mathrm{r}')/I_\mathrm{r}'$ is approximately 1×10^{-4} at the peak of the action potential. Introducing these values together with the value of P_r given above into Eq. 10 we obtain

$$P_\mathrm{a}-P_\mathrm{r} = -0.25\times10^{-4}$$

for a crab nerve stained with Pyronin B. The optical signals and, hence, the variations in polarization are very small. However, since only a small fraction of the cellular elements in the nerve is involved in nerve excitation, the observed magnitude of the optical signals does not necessarily indicate that the structural change is very small. The following calculation may serve to clarify this point.

The thickness of the major diffusion-barrier in the axonal membrane is less than 10 mμ. Individual axons are covered with layers of Schwann cells and connective tissue. There are some fibrous materials which are stained by the dye in the intercellular space. Therefore, the amount of dye molecules within the major diffusion-barrier is probably 10^{-3} to 10^{-4} times the total amount of dye molecules in the nerve. There is no doubt that the optical signals shown in Fig. 8 were derived from only a very small portion of dye molecules located within the membrane.

By using the equation derived by Weber (*26*), the degree of polarization for a nerve may be written in the following form:

$$\frac{1}{P}-\frac{1}{3} = \left(\frac{1}{P_0}-\frac{1}{3}\right)\Bigg/\left\{\frac{f_1}{1+3\tau_1/\alpha_1}+\frac{f_2}{1+3\tau_2/\alpha_2}\right\} , \tag{11}$$

where P_0 is the limiting polarization, τ_1, the lifetime of the dye molecules in the axonal membrane, α_1, the relaxation time for rotational motion of the dye molecules in the membrane, and f_1 is the fraction of the dye molecules contributing to the optical signal. The symbols τ_2, α_2 and f_2

represent the corresponding quantities for the dye molecules outside the axon membrane; these quantities are assumed not to change during the process of action potential production. Since f_1 is considered to be 10^{-3} to 10^{-4} times f_2 (and $f_1 + f_2 = 1$), it is found that the observed small decrease in P during nerve excitation derives from a large increase in τ_1/α_1 associated with production of an action potential. The optical signal observed with Pyronin B is negative; therefore, the lifetime, τ_1, must either decrease or remain unchanged during nerve excitation. Thus, it is inferred that there is during nerve excitation a marked decrease in the relaxation time, α_1, for rotational Brownian motion of the dye molecules in the membrane.

The order of magnitude of τ_1 is 10^{-9} sec (26, 28), and P_0 is known to be very close to 0.5 for Pyronin B (27). The relaxation time for rotational diffusion of a fluorescent oscillator in the resting axonal membrane is considered to be on the same order of magnitude as that in a concentrated polymer solution; therefore, it may be reasonable to assume α_1 is of the order of magnitude of 10^{-6} sec (29). If it is assumed that the optical signal in nerves stained with Pyronin B is due to changes in static quenching (27), τ_1 would remain unaltered during the action potential; the observed decrease in P is attributed in this case solely to a marked decrease in α_1, namely, to a large increase in the rotational motion of the intramembrane dye molecules during the action potential. If τ_1 decreases during an action potential, a greater reduction in α_1 is expected.

It is well known (see Secs. IV and V) that the mobilities of ions in the axonal membrane drastically increase during an action potential. This increase is related, by virtue of Walden's rule (30), to a fall in the viscosity of the media surrounding individual ions. The fall in the degree of polarization of the fluorescent light discussed in this section may be regarded as an additional piece of evidence to indicate a fall in the intramembrane viscosity during the process of nerve excitation.

XI. CONCLUSION

The development of new techniques to study the electrical and optical properties of nerve membranes has helped to elucidate the molecular events occurring in the membrane during nerve excitation. The results obtained by the use of intracellular perfusion of squid giant axons have led to a better understanding of the ionic environment that is required for the production of action potentials. Axons are found to maintain excitability in external media completely devoid of univalent cations. All-or-none action potentials can be elicited in axons with Na-phosphate internally and $CaCl_2$ as the sole external electrolyte. From these and other results, it is proposed that a conformational change in the nerve membrane macromolecules, caused by cooperative univalent and divalent cation-exchange at fixed negative sites within the membrane, is the primary event in nerve excitation.

Changes in turbidity and birefringence and changes in the intensity of fluorescence of stained nerves have been detected during excitation. These changes in optical properties may reflect a conformational change in nerve membrane macromolecules coincident with the nerve impulse.

ACKNOWLEDGEMENT

We wish to thank Dr. Laurence D. Carnay and Dr. Jörgen Fex for their valuable assistance in preparing this manuscript.

REFERENCES

1 A. L. Hodgkin, " The Conductance of the Nervous Impulse," Charles C. Thomas, Springfield, Ill., (1964).

2 T. Oikawa, C. S. Spyropoulos, I. Tasaki and T. Theorell, *Acta Physiol. Scand.*, **52**, 195 (1961).

3 P. F. Baker, A. L. Hodgkin and T. I. Shaw, *Nature*, **190**, 885 (1961).

4 L. B. Cohen, R. D. Keynes and B. Hille, *Nature*, **218**, 438 (1968).

5 I. Tasaki, A. Watanabe, R. Sandlin and L. Carnay, *Proc. Natl. Acad. Sci. U.S.*, **61**, 883 (1968).

6 J. V. Howarth, R. D. Keynes and J. M. Ritchie, *J. Physiol. (London)*, **194**, 745 (1968).

7 A. Watanabe, I. Tasaki and L. Lerman, *Proc. Natl. Acad. Sci. U.S.*, **58**, 2246 (1967).

8 I. Tasaki, I. Singer and T. Takenaka, *J. Gen. Physiol.*, **48**, 1095 (1965).

9 A. Voët, *Chem. Rev.*, **20**, 169 (1937).

10 H. G. Bungenberg de Jong, in " Colloid Science," ed. by H. R. Kruyt, Elsevier, New York, Vol. II, p. 259 (1949).

11 I. Tasaki, I. Singer and A. Watanabe, *J. Gen. Physiol.*, **50**, 989 (1967).

12 Y. Kobatake and I. Tasaki, in " Nerve Excitation," ed. by I. Tasaki, Appendix 2, Charles C. Thomas, Springfield, Ill., p. 181 (1968).

13 I. Tasaki and K. Kobatake, in " Nerve Excitation," ed. by I. Tasaki, Appendix 1, Charles C. Thomas, Springfield, Ill., p. 167 (1968).

14 I. Tasaki, A. Watanabe and L. Lerman, *Am. J. Physiol.*, **213**, 1465 (1967).

15 I. Tasaki, T. Takenaka and S. Yamagishi, *Am. J. Physiol.*, **215**, 152 (1968).

16 P. F. Baker, M. P. Blaustein, R. D. Keynes, J. Manil, T. I. Shaw and R. A. Steinhardt, *J. Physiol. (London)*, **200**, 459 (1969).

17 I. Tasaki and S. Hagiwara, *J. Gen. Physiol.*, **40**, 859 (1957).

18 J. R. Segal, *Nature (London)*, **182**, 1370 (1958).

19 C. S. Spyropoulos, *Am. J. Physiol.*, **200**, 203 ((1961).

20 I. Tasaki, "Nerve Excitation," Charles C. Thomas, Springfield, Ill., Chap. 7 (1968).

21 J. P. Changeaux, J. Thiery, Y. Tung and C. Kittel, *Proc. Natl. Acad. Sci. U.S.*, **57**, 335 (1967).

22 A. L. Lehninger, *Proc. Natl. Acad. Sci. U.S.*, **60**, 1069 (1968).

23 R. M. Barrer and J. D. Falconer, *Proc. Roy. Soc.*, A **236**, 227 (1956).

24 H. S. Sherry, *J. Phys. Chem.*, **72**, 4086 (1968).

25 I. Tasaki, L. Carnay and A. Watanabe, *Proc. Natl. Acad. Sci. U.S.*, **64**, 1362 (1969).

26 G. Weber, *Biochem. J.*, **51**, 145, 155 (1952).

27 T. Förster, in " Fluoreszenz Organischer Vergindungen," Göttingen, Vandenhoeck und Ruprecht, (1951).

28 R. F. Steiner and H. Edelhoch, *Chem. Rev.*, **62**, 457 (1962).

29 T. Yanagida, A. Teramoto and H. Fujita, *J. Phys. Chem.*, **72**, 1265 (1968).

30 R. A. Robinson and R. H. Stokes, in "Electrolyte Solution," Butterworth, London, p. 130(1959).

Received for publication, April 1, 1970.

Advan. in Biophys., Vol. 2, pp. 33–76 (1971)

ONE-ELECTRON AND TWO-ELECTRON TRANSFER MECHANISMS IN ENZYMIC OXIDATION-REDUCTION REACTIONS*

ISAO YAMAZAKI

Biophysics Division, Research Institute of Applied Electricity, Hokkaido University, Sapporo, Japan

In recent years, there have been many discussions regarding the question whether oxidation-reduction reactions in biological systems proceed by way of a one-electron or by way of a two-electron transfer. The concept of one-electron transfer processes in its general form was first put forward by Haber (*1*), and it was also suggested by Haber and Willstätter (*2*) that many oxidation-reduction reactions occur by way of a one-electron transfer. On the basis of his data on the potentiometric titration of dyestuffs Michaelis (*3*) published in 1932 his general "Theory of the Reversible Two-Step Oxidation," in which he developed the various mathematical deductions possible from the potentiometric titration curves. Since almost all stable organic compounds contain an

---

* The code number, systematic name and trivial name (besides the name used here) of enzymes which appear in this paper are as follows:

    Alcohol dehydrogenase: EC 1.1.1.1, Alcohol: NAD oxidoreductase.

    Aldehyde oxidase: EC 1.2.3.1, Aldehyde: oxygen oxidoreductase.

    D-Amino acid oxidase: EC 1.4.3.3, D-Amino-acid: oxygen oxidoreductase (deaminating).

(continued)

even number of electrons, his postulate clearly implies that the reactions are of a free radical nature. Michaelis (*4*) subsequently extended these deductions and introduced the concept of the normal potential of each univalent step.

Interest in the oxidation-reduction process of flavin compounds was also stimulated by the observation of Kuhn and Wagner-Jauregg (*5*) that a red intermediate is formed when riboflavin is reduced in an acid solution. In 1937, observing a red intermediate of an old yellow enzyme when the enzyme was partly reduced, Haas (*6*) presented the first experimental evidence that such a semiquinone might be formed in biological oxidation. However, this observation, only briefly reported without detail, has long remained isolated and has not been reproduced by others until recently (*7*). Using an ESR technique, Beinert and Sands (*8*) obtained direct evidence that free radicals are readily formed from oxidation-reduction of flavins and flavoproteins.

Another striking demonstration of the one-electron process in the

---

L-Amino acid oxidase: EC 1.4.3.2, L-Amino-acid: oxygen oxidoreductase (deaminating).

Ascorbate oxidase: EC 1.10.3.3, L-Ascorbate: oxygen oxidoreductase.

Cytochrome b_5 reductase: EC 1.6.2.2, Reduced-NAD: ferricytochrome b_5 oxidoreductase.

Cytochrome oxidase: EC 1.9.3.1, Ferrocytochrome c: oxygen oxidoreductase.

Dihydroorotate dehydrogenase: EC 1.3.3.1, L-4,5-Dihydroorotate: oxygen oxidoreductase.

Ferredoxin-NADP reductase: EC 1.6.99.4, Reduced-NADP: ferredoxin oxidoreductase.

Glucose oxidase: EC 1.1.3.4, β-D-Glucose: oxygen oxidoreductase.

Glycollate oxidase: EC 1.1.3.1, Glycollate: oxygen oxidoreductase.

Laccase: EC 1.10.3.2, p-Diphenol: oxygen oxidoreductase, p-Diphenol oxidase or urushiol oxidase.

Lactate oxidase: EC 1.1.3.2, L-Lactate: oxygen oxidoreductase.

NADH dehydrogenase: EC 1.6.99.3, Reduced-NAD: (acceptor) oxidoreductase.

NAD(P)H dehydrogenase: EC 1.6.99.2, Reduced-NAD(P): (acceptor) oxidoreductase, Menadione reductase or DT-diaphorase.

NADPH-cytochrome c reductase: not yet specified.

Old yellow enzyme: EC 1.6.99.1, Reduced-NADP: (acceptor) oxidoreductase, NADPH dehydrogenase.

Peroxidase: EC 1.11.1.7, Donor: hydrogen-peroxide oxidoreductase.

Tyrosinase: EC 1.10.3.1, o-Diphenol: oxygen oxidoreductase, o-Diphenol oxidase, catechol oxidase or phenolase.

Xanthine oxidase: EC 1.2.3.2, Xanthine: oxygen oxidoreductase.

enzyme reaction was provided by George (9) in 1952, who found that a red reaction intermediate of horseradish peroxidase with H_2O_2 is converted into a free enzyme by the addition of stoichiometric amounts of ferrocyanide:

Compound II (red) + Ferrocyanide

$\rightarrow$ Free peroxidase + Ferricyanide.

The stoichiometry of the analogous reaction was demonstrated by Chance in the same year (10):

Compound I (green) + Ferrocyanide $\rightarrow$ Compound II + Ferricyanide,

where Compound I is the primary reaction product of peroxidase with H_2O_2. Since in the reduction of Compound I the second step was rate determining even when a two-electron donor was used, it was suggested by Chance (11) that the reaction of Compound I with a two-electron donor does not proceed beyond the Compound II stage and that the collision of Compound II with a fresh donor molecule is required to liberate the enzyme. Consequently, it is very likely that the intermediates of donor molecules are free radicals. The direct physical evidence of free radical formation was obtained with ESR spectroscopy by Yamazaki *et al.* (12), in 1959.

With the aid of the ESR technique, a concrete concept for the one-electron transfer process in biological oxidation-reduction reactions was gained. It was, however, still difficult to draw a clear distinction between one-electron and two-electron transfer processes in oxidation-reduction reactions even though the problem has been the subject of numerous hypotheses. For instance, Westheimer (13) has proposed the following classification: if a molecule with an even number of electrons is oxidized or reduced to form another molecule with an even number of electrons, the process will be considered a two-electron transfer unless a species is produced which contains an odd number of electrons and which has a half time, during the reaction, longer than 10^{-11} sec. This essentially means that a free-radical intermediate, to be considered, must survive long enough to break out of the " solvent cage " in which it was formed, so that it can react with a compound other than those surrounding it at the moment of its creation.

The fundamental problem of electron transfer mechanism is well summarized in this short sentence. However, it is not practical to draw a distinction between one-electron and two-electron transfer mechanisms in terms of lifetime of a species that contains an odd number of electrons.

Experimentally it might be easier to clarify the problem of whether the free-radical intermediate breaks out of the "solvent cage" and is released into the medium. The details will be discussed in the present paper.

A TENTATIVE DEFINITION OF ONE-ELECTRON AND TWO-ELECTRON TRANSFER MECHANISM AT THE SURFACE OF ENZYME MOLECULES

In this paper, we shall discuss the mechanism of electron transfer from and to the molecules that contain an even number of electrons and that undergo an overall two-electron (bivalent) oxidation or reduction. The overall electron transfer reaction from a donor (D^{2-}) to an acceptor (A) can be expressed by a simple chemical equation:

$$D^{2-} + A \longrightarrow D + A^{2-}. \tag{1}$$

In some cases the reverse reaction also takes place.

Since a hydrogen atom consists of a proton plus an electron and many reactants have dissociable protons in an aqueous solution, there are variations of Eq. 1, for example: $DH^- + A \rightarrow D + AH^-$, and $DH_2 + A \rightarrow D + AH_2$. In these reactions, protons are transferred from donor to acceptor either directly or through the medium. Proton participation in the reactions will vary depending on the pH of the solution.

The number of electronic charges of the molecules in Eq. 1 is optional. When the donor is a reduced nicotinamide, the electron transfer reaction is usually formulated as follows:

$$DH + A + H^+ \longrightarrow D^+ + AH_2. \tag{2}$$

It has been established that one proton and two electrons are transferred from and to the nicotinamide adenine dinucleotide, during, the enzymic reactions.

Proton participation in the electron transfer reactions is very sophisticated, and it is said that probably a unified theoretical treatment of the electron transfer reactions is achieved without mentioning the participation of protons. In this sense, Eq. 1 can be used as a basic equation for the present discussion.

Figure 1 (I) shows the hypothetical steps of electron transfer reaction in Eq. 1. In this figure, the ideas of Westheimer (13) and Weiss (14) are schematized. From the same point of view as Westheimer, Weiss

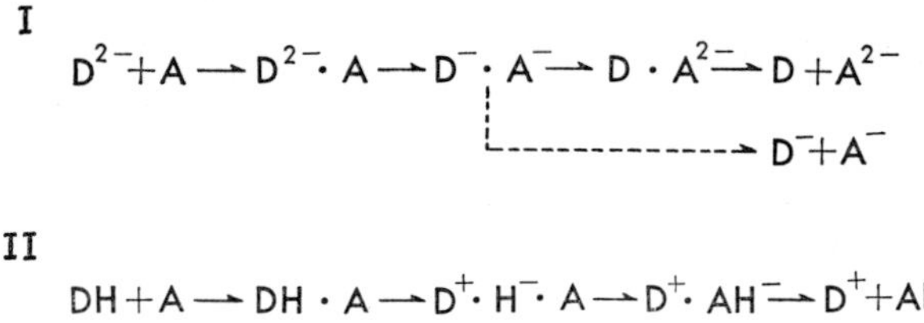

I

$$D^{2-} + A \longrightarrow D^{2-} \cdot A \longrightarrow D^{-} \cdot A^{-} \longrightarrow D \cdot A^{2-} \longrightarrow D + A^{2-}$$
$$\longrightarrow D^{-} + A^{-}$$

II

$$DH + A \longrightarrow DH \cdot A \longrightarrow D^{+} \cdot H^{-} \cdot A \longrightarrow D^{+} \cdot AH^{-} \longrightarrow D^{+} + AH^{-}$$

Fig. 1. The hypothetical mechanism of electron transfer in the overall bivalent oxidation-reduction reactions. In I, $D^{-} \cdot A^{-}$ corresponds to the primarily formed one-electron transfer intermediate that has been suggested by Westheimer (13) and Weiss (14). D^{-} and A^{-} are free radicals and react with each other to form D and A^{2-} as final products. II indicates a hydride transfer mechanism in which one proton and two electrons move simultaneously.

(14) has also suggested that, in general, the appearance of a free radical as a reacting entity depends on the intrinsic behavior of the primarily formed one-electron transfer complex ($D^{-} \cdot A^{-}$ in Fig. 1) and that the complex may release into the medium either a free radical or a valence-saturated species, the latter arising from a rearrangement within the complex prior to its breakup by the solvent (water).

Figure 1 (I) might suggest that these oxidation-reduction reactions proceed by way of one-electron changes and that two such changes may generally occur in rapid succession. In other words, the possibility that the one-electron transfer complex does exist, irrespective of its life-time, as a reaction intermediate even if it cannot be identified by the current analytical methods cannot be excluded.

This hypothetical intermediate complex appears very similar to charge transfer complexes whose physical interpretation has been proposed by Mulliken. In the case of charge-transfer complexes, a resonance interaction is assumed between a no-bond state of the isolated units and the ionic excited state of the assembly arising from the transfer of a unit charge. When reactants are both small molecules and there is orbital overlap in the complex, it is difficult to discriminate between a charge-transfer complex and a one-electron transfer intermediate. When one of two reactants is an enzyme in which the site responsible for electron transfer is localized, the one-electron transfer intermediate is considered to be a discrete chemical species, occasionally of a biradical nature. Such an intermediate has been assumed by many workers as an activated form in the reaction of enzymic oxidation-reduction, but still remains hypothetical.

Now, the question may be raised whether any reaction is known which should be classified as a typical two-electron transfer reaction in which such an intermediate complex is not formed. Westheimer (*13*) has described the addition of bromine to stilbene as an example of a polar reaction with an ionic mechanism and one that really represents a type of two-electron oxidation. Hydride transfer also belongs to this type of reaction and has been successfully chemically established. A general scheme of hydride transfer is formulated in Fig. 1(II). Hydride transfer has been suggested for the mechanism of hydrogen transfer from substrate to coenzyme or its reversal in the enzymic reaction that requires a nicotinamide derivative as coenzyme. It seems very likely, but even in this case there is still no crucial evidence for this. For these reasons, it is impossible to set up a distinction between one-electron and two-electron mechanisms in terms of existence or lifetime of the one-electron transfer intermediate.

In this paper, a tentative definition of one-electron and two-electron transfer mechanisms will be used. When the reaction proceeds along the solid lines in Fig. 1(I), it is termed a two-electron transfer mechanism regardless of the lifetime of the one-electron transfer intermediate. According to Weiss (*14*), this means that the intermediate complex releases into the medium only valence-saturated species. Of course, hydride transfer belongs to this category. A one-electron transfer mechanism then implies that the complex releases only free radicals *via* a reaction path that is indicated by the dotted line in Fig. 1(I). The free radicals thus formed will transfer one more electron between the same or different free-radical species and the final products are D and A^{2-}. This definition can be generally applied to the overall two-electron transfer reaction represented by Eqs. 1 and 2.

When such reactions are catalyzed by the enzyme that contains oxidation-reduction molecules as the prosthetic group, it is reasonable to divide the electron transfer process into two steps: from donor to enzyme and from enzyme to acceptor. Figure 2 shows a schematic illustration of electron transfer reaction from the donor (Y^{2-}) to the enzyme system and from the enzyme system to the acceptor (Y) that occur at the surface of an enzyme molecule. It should be noted here that in many cases the donor (or the acceptor) reacts with the enzyme-

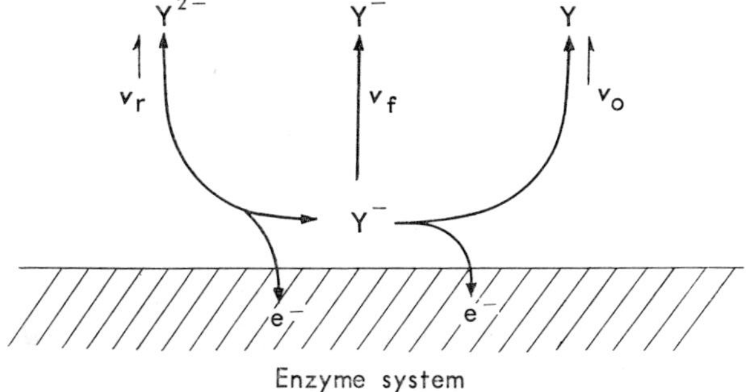

Fig. 2.   Schematic representation of electron transfer processes at the catalytic site of an enzyme.   Right and left directions indicate oxidation of the donor (Y^{2-}) and reduction of the acceptor (Y), respectively.   When Y^{2-} is oxidized by this system, the reaction velocity $(-dY^{2-}/dt)$ equals $(1/2v_f + v_o)$ after Y^- reaches a steady-state concentration.   The time is about 50 msec in the reaction shown in Fig. 3.

acceptor (or -donor) complex rather than with the enzyme itself.   By the enzyme system, either such an enzyme complex or an enzyme which is in the oxidized or reduced state is meant.   And in this broad sense, a direct electron transfer from donor to acceptor on the enzyme surface is also involved in the present discussion.

It may be appropriate to assume that the formation of free-radical intermediates takes place at the catalytic site of the enzyme in the same manner as demonstrated in Fig. 1(I).   Now, the problem is whether the intermediate is detached from the enzyme with or without a further transfer of an electron with the enzyme.   The free radical thus freed of the enzyme can be observed with the aid of the ESR technique provided that it accumulates above the level of ESR sensitivity.   The free radical is readily distinguished from that of the enzyme or of the enzyme complex by the hyperfine structure of the ESR spectrum.   In the presence of a catalytic amount of enzyme, however, ESR signals have never been detected from the enzyme.

The important problem is then, as suggested by Chance (15), that an incisive study of the role of free radicals in enzymic reactions must indicate, not only whether the free radicals are produced at detectable levels, but also whether their formation is due to the inherent catalytic activity of the enzyme.   Yamazaki and Piette (16) have introduced a parameter κ which characterizes the basic features of the enzyme-catalyzed electron transfer reactions. The κ was defined by a ratio of the velocity of free radical formation (v_f) to the velocity of donor (or acceptor) disappearance (v) at the steady state of the reaction. According to the schematic mechanism (Fig. 2), the κ will be given by the following equation:

$$\kappa = \frac{v_f}{1/2\,v_f + v_0} \qquad \text{for oxidation of } Y^{2-}, \tag{3}$$

$$\kappa = \frac{v_f}{1/2\,v_f + v_r} \qquad \text{for reduction of } Y, \tag{4}$$

if the free radicals decay only by dismutation,

$$2Y^- \xrightarrow{\;k_d\;} Y^{2-} + Y. \tag{5}$$

It is evident from the definition that the value of κ should be 0 for a typical two-electron transfer mechanism and 2 for a typical one-electron transfer mechanism. The value of κ can be obtained from experimental data as follows. In general, the decay of free radicals is fairly rapid and free radicals reach their steady state concentration within 100 msec

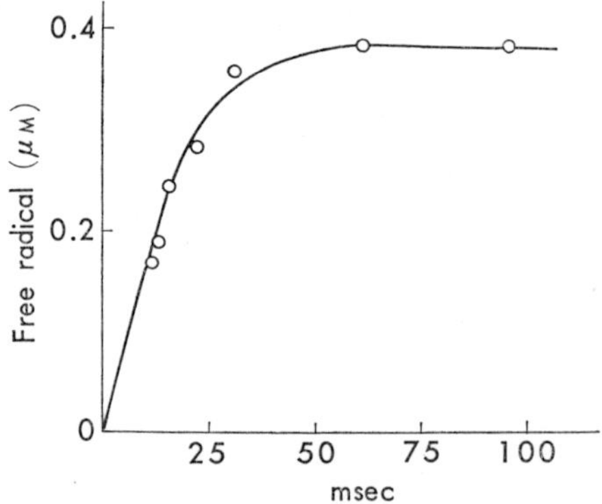

Fig. 3. Apparent free-radical formation curve in the initial stage of ascorbate oxidase reaction (Yamazaki and Piette, *16*). The magnetic field was adjusted so as to obtain the maximum of the derivative curve of ESR absorption. Each reaction time was given by the flow rate of reaction solutions. If the volume from the mixer to the center of the cavity is divided by flow rates, the quotients are reaction times.

(Fig. 3). At a steady state, the rate of free radical formation equals the rate of their decay:

$$v_f = 2k_d(Y^-)_s^2. \tag{6}$$

Then, κ will be

$$\kappa = \frac{2k_d(Y^-)_s^2}{v}, \tag{7}$$

where k_d is a rate constant of free radical dismutation and $(Y^-)_s$ is the steady state concentration of Y^-, both being estimated by use of an ESR spectrometer equipped with rapid flow apparatus (*16, 17*).

It is unambiguous that ESR spectroscopy is the most direct physical technique for detecting free radicals in solution, but the value of κ thus obtained may involve 20 or 30% errors due mainly to the inherent difficulty in flow experiments and in making quantitative estimations of free radical concentrations. Another weak point of this method is that the κ cannot be estimated when k_d is unknown or when the reaction rate is too slow to accumulate a detectable concentration of free radicals. Even in these cases, there is the possibility that the κ can be estimated. In many cases, the free radicals exhibit strong reducing activity, especially for iron complexes such as cytochromes and the iron *o*-phenanthroline complex. Consequently, it becomes possible to select experimental conditions so as to trap an unpaired electron of the free radical with a

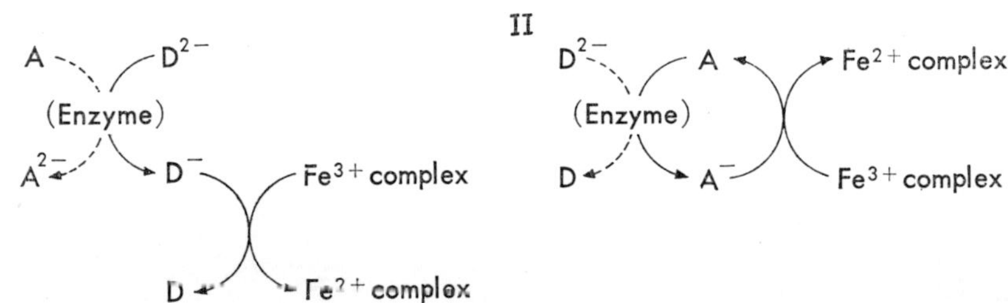

Fig. 4. The methods of κ estimation with use of free-radical scavengers. The dotted lines indicate that the reactions are two-electron transfer mechanisms. I. Oxidative formation of free radicals. The overall reaction is given in Eq. 11. The value of κ can be estimated from the mole ratio of the A or D^{2-} consumed to the iron complex reduced. II. Reductive formation of free radicals. In this reaction, A behaves as a one-electron carrier. The value of κ can be estimated using Eq. 16.

suitable one-electron acceptor.   Figure 4 shows two typical cases where such free radicals occur in enzyme reactions and decay by nonenzymic reactions with iron complexes.   When free radicals are formed oxidatively as shown in Fig. 4 (I), one-half of the donor electrons is transferred to acceptor A and the other half to the iron complex:

$$2D^{2-}+A \xrightarrow{\text{enzyme}} 2D^-+A^{2-}, \tag{8}$$

$$2D^-+2Fe^{3+} \text{ complex} \longrightarrow 2D+2Fe^{2+} \text{ complex},$$
$$\text{in the presence of } Fe^{3+} \text{ complex}, \tag{9}$$

$$2D^- \longrightarrow D^{2-}+D, \text{ in the absence of } Fe^{3+} \text{ complex}. \tag{10}$$

In the presence of the Fe^{3+} complex, the overall reaction is

$$2D^{2-}+A+2Fe^{3+} \text{ complex} \xrightarrow{\text{enzyme}} 2D+A^{2-}+2Fe^{2+} \text{ complex}. \tag{11}$$

In the case of Fig. 4(II) where free radicals are formed by accepting an electron, acceptor A behaves as a one-electron carrier:

$$D^{2-}+2A \xrightarrow{\text{enzyme}} D+2A^-, \tag{12}$$

$$2A^-+2Fe^{3+} \text{ complex} \longrightarrow 2A+2Fe^{2+} \text{ complex},$$
$$\text{in the presence of } Fe^{3+} \text{ complex}. \tag{13}$$

$$2A^- \longrightarrow A^{2-}+A, \text{ in the absence of } Fe^{3+} \text{ complex}. \tag{14}$$

In the presence of Fe^{3+} complex, the overall reaction is then

$$D^{2-}+2Fe^{3+} \text{ complex} \xrightarrow{\text{enzyme}+A} D+2Fe^{2+} \text{ complex}. \tag{15}$$

Now, the value of κ can be estimated by the analysis of reactions at a steady state:

$$\kappa = \frac{\text{rate of } Fe^{3+} \text{ complex reduction}}{\text{rate of overall reaction in the absence of } Fe^{3+} \text{ complex}}. \tag{16}$$

It is obvious that the value of κ thus obtained has a definite physical meaning when the reduction of the Fe^{3+} complex is caused only by the free radicals and the decay of free radicals by dismutation is negligibly small.   Of course, care should be used to select a free radical scavenger which does not affect the enzyme reactions.   It has been found that this method can be applied to many enzyme reactions (18–22).   The reason for this is probably due mostly to the characteristic features of enzyme reactions in which only structurally specific molecules undergo a rapid chemical conversion. Although the parameter κ can be, in principle, used for the analysis of direct electron transfer processes between donor and acceptor as shown in Fig. 1, the estimation of κ may be difficult in such nonenzymic reactions.

METAL ENZYMES (OXIDASES AND PEROXIDASES)

In 1934, Haber and Weiss (23) suggested that there was a free radical mechanism in the iron-catalyzed decomposition of hydrogen peroxide. It is also well known that trace amounts of salts from heavy metals, such as Fe, Co, Mn, Cu, *etc.*, often have profound effects upon the rates of radical chain autoxidations of organic molecules. Consequently, it is probably natural that the early applications of ESR spectroscopy in the identification of free radical formation were made mostly in the field of oxidases containing these heavy metals.

1. *Horseradish peroxidase*

Peroxidase is a hemoprotein which catalyzes the coupled oxidoreduction of hydrogen peroxide and the hydrogen donor:

$$DH_2 + H_2O_2 \xrightarrow{\text{peroxidase}} D + 2H_2O. \tag{17}$$

On the basis of information obtained experimentally by George (9, 24), Chance (10) and Yamazaki *et al.* (12), the mechanism of compulsory transformation of bivalent donor molecules to free radicals that is shown in Fig. 5 seems probable. Figure 6 shows an ESR spectrum of p-benzosemiquinone that occurs during the oxidation of hydroquinone catalyzed by peroxidase. Since the p-benzosemiquinone occurs in excess of the enzyme present and has a typical hyperfine structure, it seems very likely that such intermediate radicals are free in solution. The problem still remains unsettled as to whether or not the formation of p-benzosemiquinone is compulsory. Using Eq. 7, Yamazaki and Piette (16) have obtained the κ value of 1.3–2.0 for the peroxidase reaction when ascorbate is used as the electron donor. From this result, it is unambiguous that the free radical formation is not a side reaction but a main reaction. It is, however, still impossible to conclude that the reaction consists of only one-electron process ($\kappa=2$). The accurate value of κ

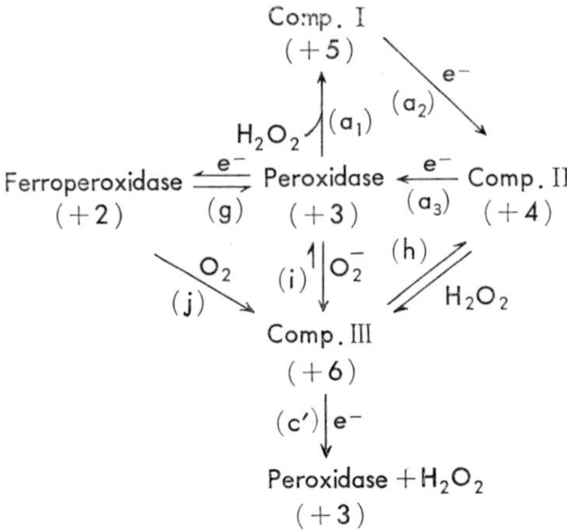

Fig. 5. Peroxidase mechanism and five oxidation forms of horseradish peroxidase. Numbers in parentheses show the effective oxidation level of the iron of peroxidase derivatives. A complete peroxidase cycle is composed of Reactions a_1, a_2 and a_3. One cycle of the reaction produces two molecules of donor free radicals.

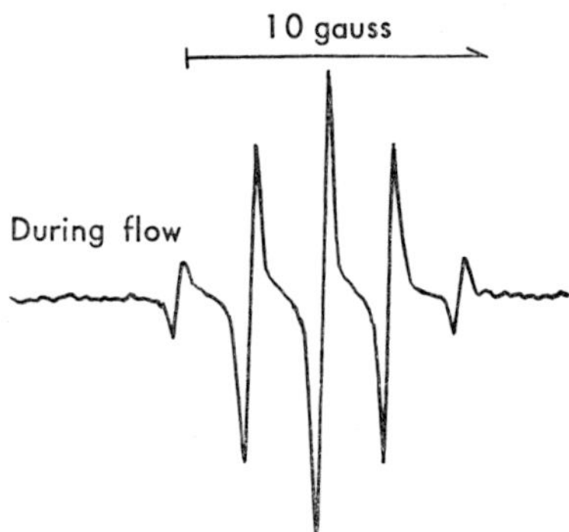

Fig. 6. The ESR spectrum of p-benzosemiquinone formed in the steady state of the peroxidase reaction during a continuous slow flow (Ohnishi *et al.* 20). The reaction time was about 70 msec and the concentration of p-benzosemiquinone was 1.2 μM. A trace amount of p-benzosemiquinone remained after the flow stopped, which was an equilibrium amount of p-benzosemiquinone that appeared in the presence of benzohydroquinone (0.5 mM) and p-benzoquinone (0.5 mM) at pH 6.5. The reaction was started in the presence of 0.1 μM enzyme, 1 mM benzohydroquinone and 0.5 mM H_2O_2.

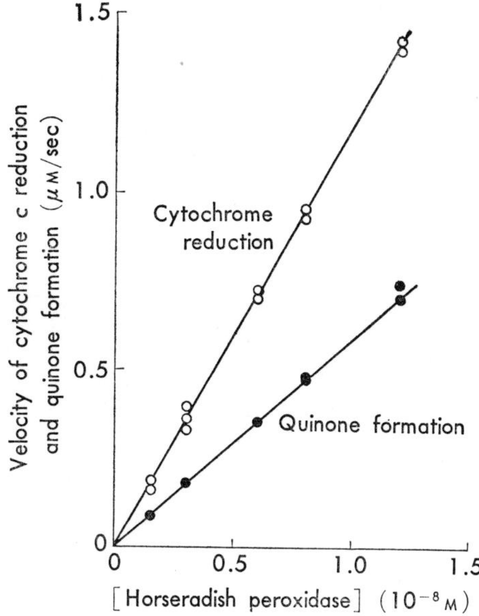

Fig. 7.   The relationship between the rates of *p*-benzoquinone formation and of cytochrome *c* reduction in the presence of various amounts of horseradish peroxidase (Ohnishi *et al. 20*).   The rate of cytochrome *c* reduction was measured in the presence of a large enough amount of cytochrome *c* (35 μM) as a scavenger of *p*-benzosemiquinone.   The rate of *p*-benzoquinone formation (overall reaction rate) was measured under the same conditions as above except that cytochrome *c* was omitted.   The ratio was estimated to be almost 2.0.

can be estimated by a purely optical method using Eq. 16.   Figure 7 shows the relation between the initial rates of *p*-benzoquinone formation in the absence of cytochrome *c* and those of cytochrome *c* reduction in the presence of cytochrome *c*.   The ratio remains almost 2.0 at various peroxidase concentrations.   The ratio thus obtained equals κ.

NADH and indoleacetate are also substrates for the peroxidase reaction, but it has been unsuccessful to observe ESR signals during the peroxidatic oxidation of these substrates.   This may be mostly due to the rapid dismutation of these free radicals and partly due to the slow reaction of these donor molecules with peroxidase-H_2O_2 compounds. Even in these cases, the stoichiometric formation of such intermediate free radicals is strongly supported by the data illustrated in Fig. 8.

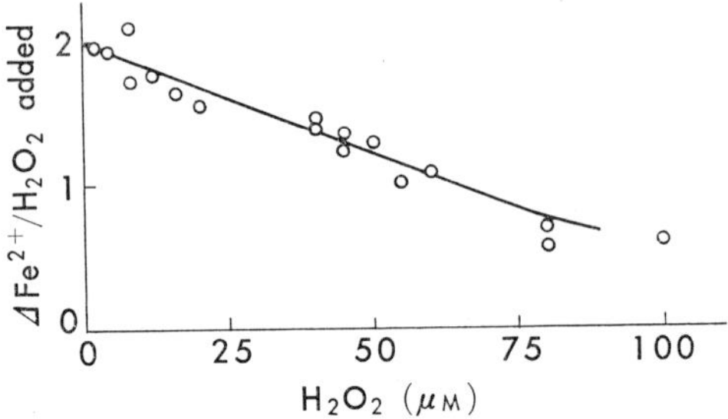

Fig. 8. The stoichiometry of iron reduction which occurs during the peroxidatic oxidation of NADH (Ohnishi *et al. 20*). The molar ratio of reduced iron
to added H_2O_2 was plotted against the concentration of added H_2O_2. The
direct reduction of iron by NADH was negligibly slow under these conditions.

Instead of cytochrome *c*, the *o*-phenanthroline iron complex is used as
the final electron acceptor. The molar ratio of iron reduced to H_2O_2
added becomes 2.0 when the H_2O_2 concentration is extrapolated to zero.
Since the direct reduction of iron by NADH is negligibly slow, the
results suggest the following mechanism:

$$2NADH + H_2O_2 \xrightarrow{\text{peroxidase}} 2NAD + 2H_2O \,, \tag{18}$$

$$2NAD + 2Fe^{3+} \text{ complex} \longrightarrow 2NAD^+ + 2Fe^{2+} \text{ complex} \,. \tag{19}$$

A similar result has been obtained for the reaction system containing
indoleacetate instead of NADH (*18*). Though the method is indirect
these results indicate that κ equals almost 2.0 when NADH and
indoleacetate are oxidized by the peroxidase system.

The next problem in the classification of electron transfer reactions
is whether H_2O_2 is reduced to water by way of a one-electron or by way
of a two-electron transfer. It has been concluded from the results of
George (*9, 24*) and Chance (*10*) that the peroxidase H_2O_2 complex is
reduced to the free enzyme in two steps of the one-electron process.
However, this argument does not necessarily apply to the reduction of
H_2O_2. In this respect, it is of interest to note that the iron of the
intermediate enzyme (Compound II) is thought to be in a ferryl form by
many workers (*25–27*), especially recently by Maeda *et al.* (*28, 29*) from
the Mössbauer effect. They also suggested that the iron of Compound
I seems to be in a ferryl form. It may be concluded that the two
oxidizing equivalents of H_2O_2 are divided into two chemical species in

the state of Compound I as in the case of cytochrome c peroxidase (*30*). However, it might be still impossible to conclude that the transfer of two valence electrons to H_2O_2 is not " simultaneous." The purpose of the present paper is not to discuss the details of the electron transfer mechanism within the intermediary complex between donor, acceptor and enzyme but to discuss the nature of the primary reaction products of donor or acceptor molecules from the point of view of valence saturation. In this respect there is no doubt about the fact that the primary product of H_2O_2 reduction is a valence-saturated species, but not a free radical. Consequently, according to the present definition, the mechanism of H_2O_2 reduction in the peroxidase system is termed a two-electron transfer mechanism.

2.   *Laccase and ascorbate oxidase*

Both enzymes are blue copper proteins which catalyze the following reactions:

$$2\ \text{Hydroquinone} + O_2 \xrightarrow{\text{laccase}} 2\ p\text{-Benzoquinone} + 2H_2O . \qquad (20)$$

$$2\ \text{Ascorbate} + O_2 \xrightarrow{\text{ascorbate oxidase}} 2\ \text{Dehydroascorbate} + 2H_2O . \quad (21)$$

The copper of these enzymes is considered to be in the cupric form. It is generally accepted that the copper is reduced by substrates and reoxidized by O_2.

Using an ESR spectrometer equipped with continuous flow apparatus, Nakamura (*31, 32*) observed the formation of p-benzosemiquinone during the laccase-catalyzed oxidation of benzohydroquinone. This was confirmed in an elaborate study by Broman *et al.* (*33*), who observed, during the catalytic action of fungal laccase and human ceruloplasmin, the rise and decay of a free-radical intermediate of p-phenylenediamine, a substrate specifically chosen for this purpose. A similar study was made by Ogura and Nakamura with catechol oxidation by laccase (*34*). Yamazaki and Piette (*16*) observed the ESR absorption spectra of monodehydroforms of ascorbate and reductate during their oxidation in the presence of ascorbate oxidase systems. Figure 9 shows the spectra of monodehydroascorbate given by a recent estimation, using the different modulation amplitude of the magnetic field (*20*). The resolution in the spectrum of Fig. 9C is still somewhat lower than that obtained by Foester *et al.* (*35*) at the equilibration of fully oxidized and fully reduced

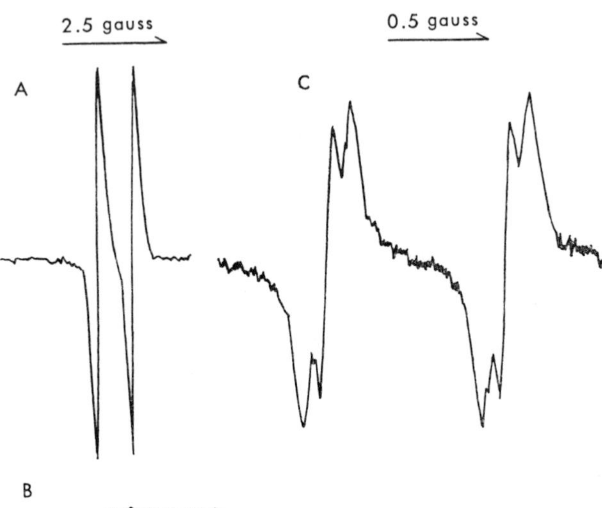

Fig. 9.   The ESR spectra of the ascorbate free radical formed in the steady state of ascorbate oxidase reaction during a continuous slow flow (Ohnishi *et al.* *20*).   The enzyme was 1 μM (on the basis of copper) and the free radical was 2 μM.   A and C were taken during the flow and B after the flow stopped.   The amplitude of field modulation was 0.1 gauss in A and B and 0.04 gauss in C. The gain used in C was twice that in A and B.

forms of ascorbate.   This seems to be partly due to the inevitable disturbance of using a flow method in the experiment in Fig. 9C.

The value of κ in the oxidation of ascorbate by the ascorbate oxidase system was very roughly estimated to be 1.4–2.2 by use of Eq. 7 (*16*). The more reliable value close to 2 was given by a study using cytochrome *c* as a scavenger of the monodehydroascorbate formed during the reaction (*19*).   The " $\kappa=2$ mechanism " might be simply explained by assuming that each electron donor reacts with the single cupric ion present on the enzyme surface and is detached from the enzyme after it transfers its single electron to the cupric.   Since these enzymes contain more than 4 cupric ions per each molecule, it could be concluded that there is little cooperative function between the cupric ions when they accept electrons from the donor molecules.

Another characteristic feature in the mechanism of these oxidase reactions is that O_2 is reduced to water without any appreciable formation of intermediates (*36*).   For instance, the formation of H_2O_2 as an intermediate can be excluded by the fact that H_2O_2 is neither produced nor consumed by the reaction. This fact also excludes a possible formation of the superoxide radical anion (O_2^-) because its formation inevitably results in the formation of H_2O_2. Roughly speaking, two mechanisms can be assumed: first, O_2 is detached from the enzyme after it receives 4 electrons; second, each atom of O_2 is detached from the enzyme separately after an oxygen atom receives 2 electrons. The abscission of the oxygen bond will occur at a certain step of the successive electron transfer from enzyme to O_2. This probably occurs at the third electron step in solution but it may not necessarily be applied to the reaction on the enzyme surface. Although it is very unlikely that 4 electrons move simultaneously to O_2, the main reaction of O_2 reduction by the ascorbate oxidase system is to be classified as a four-electron transfer mechanism rather than a two-electron transfer one on the basis of the analysis of its primary product. A similar discussion has described the mechanism of H_2O_2 reduction in the peroxidase reaction. There must be an elaborate mechanism in these oxidases that enables them to catalyze the one-electron oxidation of two-electron donors and four-electron reduction of O_2 in the same catalytic reaction.

3. Tyrosinase

Tyrosinase is known to be a cuprous enzyme that has complex catalytic functions. It catalyzes hydroxylation of monohydroxybenzene derivatives and dehydrogenation of catechol derivatives. Judging from the overall reaction the latter reaction is very similar to the laccase reaction:

$$2 \; \text{(catechol)} + O_2 \xrightarrow{\text{tyrosinase}} 2 \; \text{(o-benzoquinone)} + 2H_2O \; . \tag{22}$$

By analyzing the primary reaction product in the catechol oxidation with ESR spectroscopy, Mason *et al.* (*37*) found that the mechanism of Reaction 22 is quite different from that of the laccase reaction. They concluded that 94% or more of the primary products of the tyrosinase-catalyzed oxidation of catechol is *o*-benzoquinone, the two-electron oxidized form of catechol. The conclusion that the reaction involves a two-electron transfer mechanism can be reached only after careful consideration. Of course, the inability to detect the intermediate free radicals with ESR spectroscopy does not always mean that the primary product is a valence-saturated molecule. One of the most convenient methods for clarifying this point is to compare it with the standard reaction in which the same electron donor is oxidized by a typical one-electron transfer mechanism ($\kappa = 2$). The peroxidase reaction may be the best one for this purpose because many electron donors can be oxidized by H_2O_2 in the presence of a relatively small amount of this enzyme and the mechanism of the peroxidase reaction is well known. During the enzymic oxidation of catechol at the same pH (5.3) and at a similar rate, the steady state concentration of o-benzosemiquinone was found to be 0.4 μM in the peroxidase reaction, but below the level of detectability by the ESR spectrometer (0.1 μM) in the tyrosinase reaction. These observations led to the conclusion that the predominate primary product of catechol in the tyrosinase reaction is o-benzoquinone and that κ is close to zero in the oxidation of catechol by tyrosinase system.

Little is known about the process of O_2 reduction by tyrosinase. Hamilton has discussed the mechanism in his recent review (*38*) and proposed the " oxene mechanism " by which an oxygen atom is transferred to the substrate in its hydroxylation reaction.   The process of the reduction of O_2 to water in the reactions of tyrosinase and laccase will also differ because of the remarkable difference in their copper form. In Reaction 22, however, H_2O_2 is neither formed nor consumed as in the case of laccase and ascorbate oxidase.   Consequently, the mechanism again falls into the four-electron transfer category when O_2 is reduced to water in the tyrosinase reaction.

FLAVOPROTEINS

The flavoprotein has a bivalent oxidation-reduction molecule in its catalytic site and the flavin molecule is reduced by suitable electron donors.   The information gained in recent years on the intermediate of flavoproteins that occurs during their oxidation-reduction has been reviewed by Beinert and Sands (*8*) and Beinert and Palmer (*39*).   In addition to the early discovery of a red intermediate of the old yellow enzyme by Haas (*6*), a large number of flavoproteins have been found to form stable semiquinoid intermediates during oxidation and reduction.   Of course, ESR has been the most powerful method for the identification.   Reliable data which indicate the occurrence of flavin free radicals in a meaningful amount have been reported mostly in metal flavoproteins.   The correlation of optical and ESR data on semiquinone formation had been reported in various flavoproteins and the results were summarized by Beinert and Palmer (*39*) in the statement that only one metal-free flavoprotein among many examined thus far, shows, upon the addition of a substrate, a free radical signal representing a significant proportion of the flavin, while, nevertheless, deeply colored intermediates appear in most cases. The exception was microsomal NADPH-cytochrome *c* reductase (*40*), but it was later concluded that the semiquinone represents only 7% of the total flavin present (*41*). A similar discussion was also made about D-amino acid oxidase (*42*). D-Amino acid oxidase forms a red intermediate with an ESR signal when the enzyme is partly reduced by dithionite, but its active intermediate observed during catalytic reaction is a purple one without an ESR signal. The purple complex is converted to a red semiquinone form by aging or illumination.

Many metal flavoproteins have been found to form ESR-detectable free-radical intermediates of the flavin type that show quantitatively and kinetically the expected behavior (*39*). However, little has been reported on the ESR study of free radicals derived from the substrates until quite recently. The following discussion will be mostly concerned with free radicals of acceptors that appear during the reactions catalyzed by flavoproteins.

1. *Flavoproteins in electron transport systems*

A characteristic feature of these flavoproteins is that they are present in subcellular particles, such as mitochondria, microsomes and chloroplasts. NADH dehydrogenase and succinic dehydrogenase are in mitochondria, cytochrome b_5 reductase and NADPH-cytochrome *c* reductase are in microsomes and ferredoxin-NADP reductase is in chloroplasts. These enzymes are known to participate in the electron-transfer reactions in particles and can be solubilized by the proper methods.

In addition to their physiological electron acceptors, the solubilized enzymes are able to catalyze rapid electron transfer from their donors to ferricyanide. It is now of special interest to clarify whether the enzymes catalyze a one-electron reduction or a two-electron reduction when two-electron acceptors are reduced by each enzyme system. Contrary to the strict specificity for the electron donors, these enzymes can catalyze the reduction of a number of electron acceptors. Among many two-electron acceptors, *p*-benzoquinone seems to be the best acceptor for the analysis of the electron transfer mechanism since it is reduced by each of these enzyme systems at a fairly high rate and the nature of

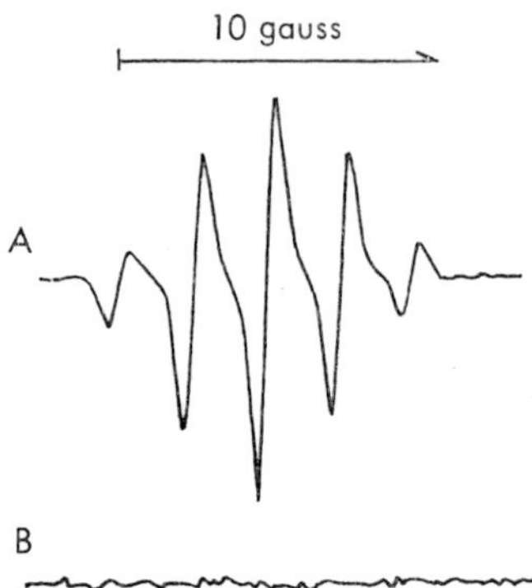

Fig. 10. The ESR spectrum of p-benzosemiquinone (A) formed in the steady state of cytochrome b_5 reductase reaction during a continuous slow flow (Iyanagi and Yamazaki, *68*). The enzyme was 0.5 μM and p-benzosemiquinone was 0.8 μM. The same magnetic field was scanned in B soon after the flow stopped.

TABLE I

Estimation of κ by ESR and Optical Spectroscopic Methods in Reactions of p-Benzoquinone Reduction Catalyzed by Flavoproteins

Reducing systems	ESR			Optical[b]		
	Rate of quinone reduction (μM sec^{-1})	Semiquinone concn. at steady state (μM)	κ[a]	Rate of cytochrome c reduction (μM sec^{-1})	Rate of quinone reduction (μM sec^{-1})	κ[c]
NADH and cytochrome b_5 reductase	14.3	0.47	2.2	1.18	0.59	2.00
NADH and NADH dehydrogenase	13.8	0.46	2.1	0.70	0.36	1.95
NADPH and ferredoxin-NADP reductase				0.90	0.46	1.96
NADH and NAD(P)H dehydrogenase	13.2	<0.02	<0.004	0.00	0.71	0.00

Reactions were carried out at 25°C (*22*).   [a] The values of κ were calculated from Eq. 7 assuming that k_d is 7×10^7 M^{-1} sec^{-1} (*20*).   [b] Enzyme concentrations used in these reactions are no more than several percent of those used in ESR experiments.   [c] The values of κ were calculated from Eq. 16.

p-benzosemiquinone which is the product of its one-electron reduction
has been well characterized (*17, 20*).

Figure 10 shows the ESR spectrum of p-benzosemiquinone which
appears during the reduction of p-benzoquinone in the presence of
cytochrome b_5 reductase and NADH. The spectrum is the same as
that observed in the peroxidase reaction (Fig. 6). The reaction is of the
first-order with p-benzoquinone in the case of cytochrome b_5 reductase
and is close to the zero-order in the case of NADPH-cytochrome c
reductase. The difference in kinetics of the two reactions reflects well
on the steady state concentration of p-benzosemiquinone as expected
from Eq. 7 (Fig. 11). The values of κ can be estimated from these experimental results and are listed in Table I.

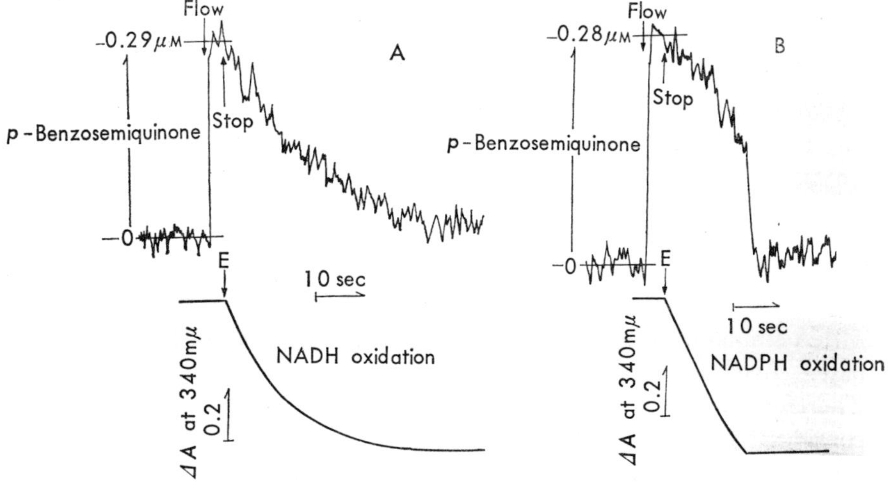

Fig. 11. Time courses of oxidation of the reduced pyridine nucleotides and
p-benzosemiquinone decay in the reactions of cytochrome b_5 reductase (A)
and NADPH-cytochrome c reductase (B) (Iyanagi and Yamazaki, *68*). Upper
diagrams show the time courses of p-benzosemiquinone concentration and
lower diagrams show the disappearance of NADH (A) and NADPH (B). The
flow in the upper diagrams was stopped almost simultaneously with the start
of the reaction start in the lower diagrams. In order to measure the p-benzo-
semiquinone concentration in the upper diagrams, the magnetic field was
adjusted so as to obtain the maximum of the derivative curve of ESR absorption. The concentrations of cytochrome b_5 reductase and NADPH-
cytochrome c reductase were 0.054 μM and 1.2 μM, respectively. The initial
rates of both reactions were almost identical.

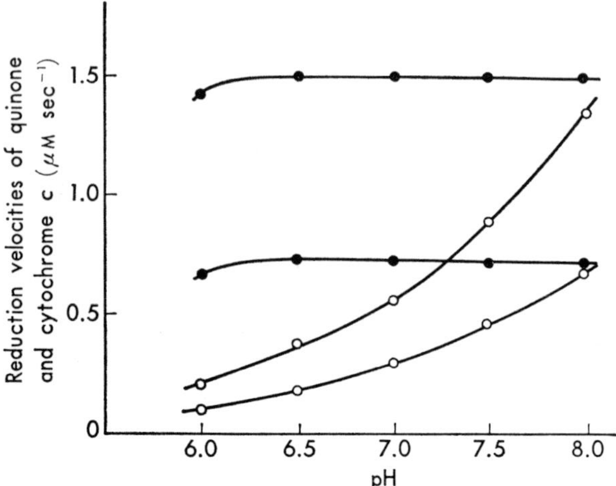

Fig. 12. The pH dependency of the rates of oxidation of reduced pyridine nucleotide (overall reaction) and the rates of cytochrome c reduction in the reactions of cytochrome b_5 reductase (black circle) and ferredoxin-NADP reductase (open circle) (Iyanagi and Yamazaki, 22). The rates of overall reaction were measured in the absence of cytochrome c. The rates of ferredoxin-NADP reductase reaction were pH dependent while the rates of cytochrome b_5 reductase reaction were almost independent of the pH tested. In both cases, the ratio of two rates (cytochrome c reduction and overall reaction) was estimated to be almost 2.0 in the range of pH between 6 and 8.

A similar experiment has been carried out with NADH dehydrogenase. According to the definition, the κ value cannot be above 2. Therefore, it is suggested that the ESR data thus obtained may involve considerable errors. When cytochrome c is used as a scavenger of p-benzosemiquinone, accurate estimations of the κ value can be made according to Eq. 16, except for the reaction of NADPH-cytochrome c reductase. The values of κ thus obtained are close to 2 and remain constant in the range of pH tested as shown in Fig. 12. This method is so reliable that it may be concluded without further confirmation by ESR methods that p-benzoquinone is reduced with the one-electron mechanism in the reaction of ferredoxin-NADP reductase. In the case of NADPH-cytochrome c reductase, a set of menadione and cytochrome b_5 is preferably used for the analysis of the mechanism of quinone reduction. Cytochrome b_5 is found to be a good electron acceptor for semiquinones of menadione. In this case, the reaction should be carried out under anaerobic conditions since the semiquinone and reduced cytochrome b_5 react with O_2.

The data listed in Table I may indicate that the mechanism of one-electron transfer to acceptor during their catalytic reaction is inherent in the flavoproteins located in electron transport systems. It was also suggested by King (*43*) and by Singer and Kearney (*44*) that succinic dehydrogenase catalyzes one-electron transfer to phenazinemethosulfate. Though their observations do not seem satisfactory for their conclusion, it is very likely that the two-electron acceptors are reduced with the one-electron mechanism in the succinic dehydrogenase reaction. It is of special importance to note that metal is unnecessary for the enzyme to function as a one-electron transfer catalyzer. The soluble flavoproteins isolated from microsomes and chloroplasts are known to be metal-free. Furthermore, it has been suggested by Hatefi *et al.* (*45*) and Pharo *et al.* (*46*) that the labile sulfide-iron system might not be involved in the diaphorase activity of the soluble NADH dehydrogenase of mitochondria.

The reactions of these enzymes with the electron donors are rather specific. As suggested by Strittmatter in the case of NADH cytochrome b_5 reductase (*47*), it might be concluded that, in general, succinate and nucleotides remain bound to the flavoprotein during the entire oxidation. Because of their specific combination between enzyme and donor, there will be little chance for the donors to be freed of the enzyme just after they transfer a single electron to the enzyme system. When metal does not participate in the reaction, it is clear that a free radical is formed as an intermediate in the enzyme or its complex with the donor after a single electron is transferred to the acceptor. The occurrence of such a free radical can be confirmed with ESR spectroscopy.

However, these free radicals have been thus far identified only in the reactions of metal flavoproteins (*39*). In this sense, it is of interest that Strittmatter (*47*) reported the appearance of a transient absorption band at 530 mμ when reduced cytochrome b_5 reductase is oxidized by ferricyanide. He suggested that the oxidation of the reduced flavoprotein-oxidized pyridine nucleotide complex appears to involve two distinguishable one-electron reactions. Whether or not the intermediate he observed is the compulsory free radical that occurs in the enzyme during its catalytic reaction, however, will have to be shown by further work. At any rate the pyridine nucleotide does not seem to be detached from the enzyme until it is fully oxidized.   Therefore, the value of κ is considered to be zero in the NADH oxidation caused by the presence of the cytochrome b_5 reductase system.   Since the free radicals of NADH are assumed to reduce O_2 (*48, 49*), the two-electron mechanism is also suggested from the results that O_2 consumption is not observed during the catalytic oxidation of NADH.

2.   *NAD(P)H dehydrogenase (DT diaphorase)*

From the point of overall reaction, NAD(P)H dehydrogenase catalyzes a reaction very similar to the mitochondrial NADH dehydrogenase:

$$NADH + Q + H^+ \xrightarrow{\text{dehydrogenase}} NAD^+ + QH_2. \tag{23}$$

The clear distinction between the two enzymes is their state of existence in biology.   NAD(P)H dehydrogenase is found in the soluble fraction of cells.   One more distinction is that NADPH is also a fast electron donor for the NAD(P)H dehydrogenase.

Similar enzymes have been isolated from many sources and sometimes are called quinone reductase (*50*).   From the data reported previously, there is some reason to predict that the NAD(P)H dehydrogenase will catalyze a two-electron reduction of the quinone.   For instance, Ernster *et al.* (*51*) have demonstrated that the enzyme can catalyze the reduction of cytochrome c in the presence of naphthoquinone derivatives, but not in the presence of benzoquinone derivatives.   It has been reported by Marki and Martius (*52*) that the mammalian quinone reductase reduces menadione very rapidly and the autoxidation of reduced menadione becomes rate-limiting after the menadione is consumed when the reaction is carried out under aerobic conditions.   A similar result has been reported by Nishibayashi *et al.* (*53*) who have shown that the amount of NADPH oxidized is equivalent to that of the menadione added under aerobic conditions.   These results are inconsistent with the mechanism that involves the formation of free radicals from acceptors and differ from those observed in the reaction of flavoprotein dehydrogenases in electron transport systems.   This distinction becomes clear from both ESR and optical data as can be seen in Figs. 13 and 14. All results are summarized in Table I.

The difference in the function as a carrier between naphthoquinone derivatives and benzoquinone derivatives that was shown in the ex-

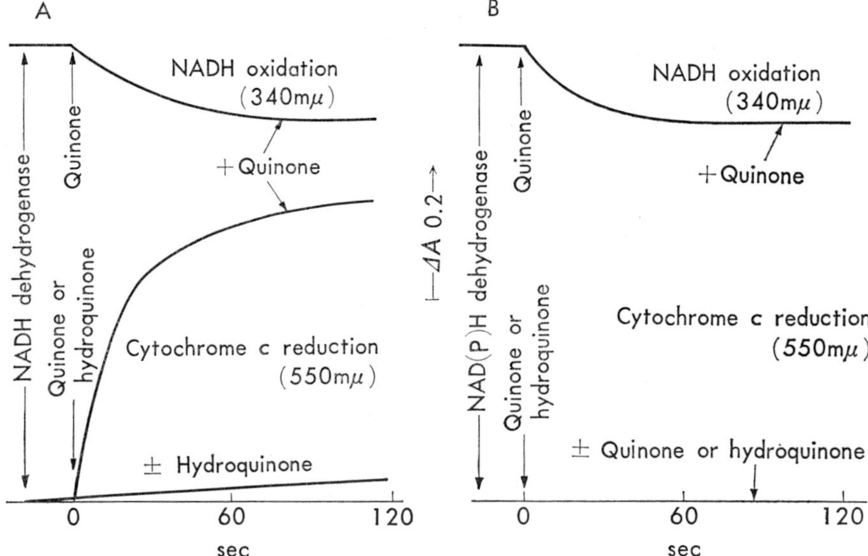

Fig. 13.   Difference in the reaction mechanism between NADH dehydrogenase (A) and NAD(P)H dehydrogenase (B).   In A, very slow reduction of cytochrome c was observed in the presence of NADH and the enzyme, which showed that cytochrome c was a slow but direct electron acceptor for the enzyme.   In both cases, the addition of benzohydroquinone did not cause the reduction of cytochrome c.   In A, p-benzoquinone induced the reduction of cytochrome c while in B, it did not.

periment of Ernster *et al.* (*51*) is ascribed to the fact that the fully reduced form of naphthoquinone derivatives is much more reactive as a reductant than that of benzoquinone derivatives and, for instance, reduces cytochrome c very rapidly.   Therefore, in order to analyze the electron-transfer mechanism with naphthoquinone derivatives as electron acceptors, cytochrome c should be replaced by cytochrome b_5 that has a low redox potential and is more resistant to reduction by naphthohydroquinone derivatives.   If the purpose is confined to a confirmation of the two-electron mechanism, the inability to consume O_2 during the reduction of naphthoquinone derivatives may be strong evidence for the conclusion, because the formation of naphthosemiquinones results in the O_2 consumption.   It is now concluded that the NAD(P)H dehydrogenase catalyzes a two-electron reduction of quinones and does not produce free semiquinones.   This might be the most characteristic

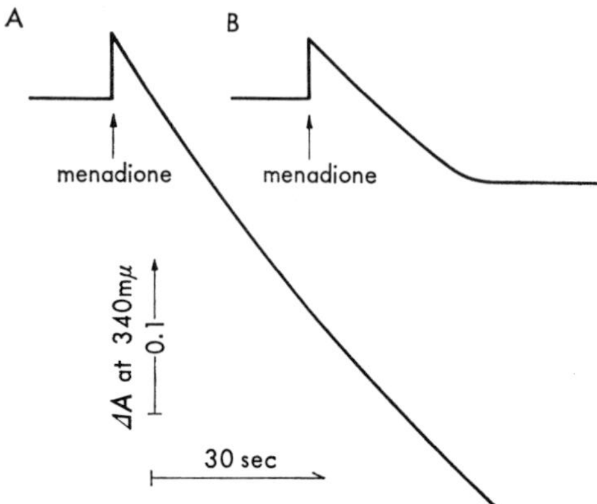

Fig. 14.   Difference in the reaction mechanism between NADH dehydrogenase (A) and NAD(P)H dehydrogenase (B) (Iyanagi and Yamazaki, *22*).
In A, the addition of menadione caused the O_2-consuming oxidation of NADH
while in B, the menadione caused the oxidation of a stoichiometric amount
of NADH equivalent to the menadione added.   In A, menadione behaved as
an electron carrier from the enzyme to O_2; while in B, it did not.

feature of the NAD(P)H dehydrogenase in comparison with the mitochondrial NADH dehydrogenase.   Because of the difficulty of purifying
the enzyme in large quantities, the reactions with pyridine nucleotides
are not yet well characterized.

3.   *Xanthine oxidase and related flavoprotein oxidases*

In addition to peroxidase, xanthine oxidase has been regarded as an
enzyme that might possibly produce free radicals from the substrate
molecule.   The possibility was suggested by Fridovich and Handler
from the results of the initiation of sulfite autoxidation (*54*) and the in-
duction of chemiluminescence (*55*), both caused by the xanthine oxidase
reaction.   Based on these results, they have suggested a formation of
oxygen radicals by which the reduction of cytochrome *c* is mediated
(*56, 57*).   Knowles *et al.* (*58*) have recently observed an ESR signal
of the oxygen radical by the rapid-freezing technique during the oxidation
of substrates by O_2 in the presence of xanthine oxidase.   However, it may
not be easy to use this method for quantitative analysis of the electron
transfer process from xanthine oxidase to O_2.

It seems to be beyond doubt that the κ is not zero when O_2 is
reduced in the xanthine oxidase reaction.   Our great interest is then to

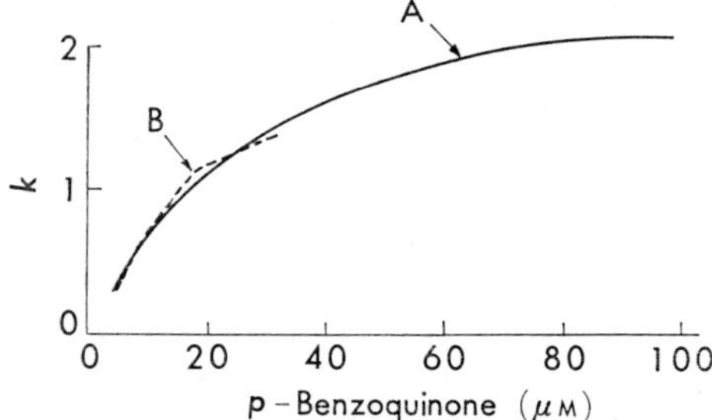

Fig. 15.   Dependence of κ value upon the p-benzoquinone concentrations in xanthine oxidase system (Nakamura and Yamazaki, *21*).   A was calculated from ESR data using Eq. 7 and B was calculated from the data of cytochrome *c* reduction using Eq. 16.

know the exact value of κ.   In the early study by Handler *et al.* (*56*) in which the rates of O_2 consumption and cytochrome *c* reduction are compared, they have suggested that the formation of oxygen radicals increases with increasing O_2 concentration while the rate of O_2 consumption remains constant.   This assumption is of special interest from the point of view that it indicates a mixed mechanism of one-electron and two-electron processes and that the mechanism may vary with the O_2 concentration.   The assumption made by Handler *et al.* (*56*) has been confirmed by ESR and optical experiments, especially using p-benzoquinone as an electron acceptor.   Figure 15 shows the dependence of the κ value upon the concentration of p-benzoquinone.   The value of κ was estimated by using either Eq. 7 or 16.   When O_2 is used as an electron acceptor, an approximate value of κ may be obtained by using the excess peroxidase as a scavenger of oxygen radicals.   Peroxidase is assumed to react with oxygen radicals forming the so-called " Peroxidase Compound III " (*48*).   Under suitable experimental conditions, the amount of Compound III accumulated may equal that of the oxygen radicals produced (*21*).   The results thus obtained (*21*) also suggest that in the xanthine oxidase reactions the mechanism of O_2 reduction is identical with that of p-benzoquinone reduction.

Xanthine oxidase is a complex enzyme which has a large molecular weight and has iron and molybdenum in addition to flavin.   Unlike the previous assumption by Handler *et al.* (*56*), Komai *et al.* (*59*) have shown that the formation of the oxygen radical requires direct electron transfer from a reduced form of flavin to O_2 in the xanthine oxidase reaction. The reaction site of xanthine has been suggested to be molybdenum by Bray *et al.* (*60, 61*), mostly based on their ESR experiments. By observing an increase in paramagnetism upon the reaction that is greater than can be explained on the basis of changes in the electronic configurations of the prosthetic groups, Ackerman and Brill (*62*) have suggested that the difference might be attributed to the formation of xanthine free radicals. However, judging from the recent ESR study of Knowles *et al.* (*58*), it seems more likely that the free radical is derived from O_2 rather than xanthine. The mechanism of the electron transfer process between xanthine and xanthine oxidase will have to be analyzed by further study.

Aldehyde oxidase is closely related to xanthine oxidase in the composition of their electron transfer apparatuses. The reaction of aldehyde oxidase with O_2 is also thought to be similar to that of xanthine oxidase (*56*). Physiologically, dihydroorotic dehydrogenase catalyzes electron transfer between dihydroorotate and NADH. This enzyme, however, also catalyzes O_2-consuming oxidation of NADH and during the reaction a considerable amount of cytochrome *c* is found to be reduced (*63*). By a rough evaluation of κ on the basis of Eq. 16, κ seems to lie between 0 and 2 and is clearly apart from both extreme cases. The old yellow enzyme also catalyzes the reduction of cytochrome *c* to a not negligible extent under aerobic conditions. A reliable method for estimating κ in the oxidase reaction will be discussed in the following section.

4. *Flavoprotein oxidases*

We shall discuss here O_2-obligative flavoprotein oxidases. The enzymes which belong to O_2-obligative flavoprotein oxidases are D- and L-amino acid oxidase, glucose oxidase, lactate oxidase and glycollate oxidase. The enzymes catalyze the following reaction:

$$DH_2 + O_2 \xrightarrow{\text{oxidase}} D + H_2O_2 . \tag{24}$$

The stoichiometry of the reaction has been confirmed in various flavoprotein oxidases. With respect to the overall kinetics, the same is the case with xanthine oxidase and the old yellow enzyme. These enzymes, however, are able to transfer electrons to the other molecules including ferricyanide. In order to analyze the electron transfer mechanism in the O_2-reduction processes, one has to detect the formation of the superoxide anion radical (O_2^-). The most direct method is an ESR method that was developed by Knowles *et al.* (*58*) in the xanthine oxidase reaction.

An important analytical method for the investigation of the participation of O_2^- in biological oxidations has recently become available through the discovery of the superoxide anion dismutase activity of erythrocuprein:

$$2O_2^- + 2H^+ \xrightarrow{\text{erythrocuprein}} H_2O_2 + O_2 . \tag{25}$$

McCord and Fridovich (*64*) found that very low concentrations of erythrocuprein inhibit the O_2-dependent reduction of cytochrome *c* catalyzed by xanthine oxidase. This method was applied by Massey *et al.* (*65*) to the analysis of flavoprotein oxidase reactions. They have concluded that a positive effect of erythrocuprein in inhibiting flavoprotein-catalyzed cytochrome *c* reduction is a strong indication of O_2^- production, while it is not possible to give the negative result such a clear-cut interpretation. They pointed out the possibility that O_2^- might indeed be produced in the reaction of the reduced flavoproteins with O_2, but the radicals are sufficiently tightly bound to the enzyme that both cytochrome *c* and erythrocuprein would be ineffective scavengers of O_2^-. According to the present definition, however, the negative result is strongly indicative of the two-electron mechanism.

DEHYDROGENASES

There are a great many enzymes that catalyze the hydrogen transfer, acting on the CH-OH groups and the aldehyde or keto-group of donors with NAD or NADP as an acceptor. These enzymes represented by a well-known alcohol dehydrogenase have no prosthetic molecule like flavins or transition metal that undergoes oxidation and reduction itself. It has been shown, in every dehydrogenase reaction thus far investigated, that hydrogen transfer to the C-4 position of the pyridine nucleotide is stereospecific, as is the hydrogen abstraction from the donor molecule. The stereochemistry is illustrated in the reversible reduction of acetaldehyde by NADH. This stereospecificity does not necessarily mean that the hydrogen transfer between donor and pyridine nucleotide is a direct hydride (H⁻) transfer process, but is indicative of the rigid constraints on the geometry of reactive centers in the ternary complex. In this respect, it is very unlikely that the free radical of either donor or acceptor is detached from the enzyme during reaction.

CLASSIFICATION OF THE ELECTRON TRANSFER MECHANISM IN THE ENZYMIC OXIDATION-REDUCTION REACTIONS ON THE BASIS OF THE FORMATION OF FREE RADICALS DERIVED FROM DONOR AND ACCEPTOR

When a molecule that undergoes one-electron oxidation and reduction, such as ferricyanide or one of the cytochromes, is oxidized or reduced in the presence of an enzyme, it can be said that the enzyme has an inherent ability to catalyze a one-electron oxidation or reduction. However, if a molecule with an even number of electrons is oxidized or reduced in the presence of the same enzyme, it does not always mean that the one-electron transfer again occurs forming free radicals of the molecule. It is the purpose of the present report to propose a way of characterizing the reactions, quantitatively if possible.

In his excellent review, Mason (*66*) has proposed the classification of oxidases in the following stoichiometric categories: (a) one-equivalent donors, two-equivalent acceptor (H_2O_2); (b) one-equivalent donors, four-equivalent acceptor (O_2); (c) two-equivalent donors, two-equivalent acceptor (O_2); and (d) two-equivalent donors, four-equivalent acceptor (O_2). The present classification is also based upon experimental results similar to Mason's.

Table II shows an amplification of Mason's classification that covers flavoprotein dehydrogenases in addition to oxidases. If the two-equivalent acceptor means that the primary product of O_2 in enzymic reduction is H_2O_2, this does not necessarily fit the case of xanthine oxidase and old yellow enzyme. On the contrary, ferricyanide is a very fast electron-acceptor in the reactions of xanthine oxidase and old yellow enzyme, but it is not pertinent to classify the enzymes into the category of the one-equivalent acceptor. It might be suggested that such a classification of enzymes should be determined using physiological substrates. The problems in Mason's classification will be: first, his classification cannot be determined until its physiological substrates are known; second, his classification is not appropriate for classifying

TABLE II

Classification of Oxidases and Dehydrogenases[a] according to Mason's Categories (66)

---

1. One-equivalent donors, two-equivalent acceptors:
    Peroxidase, ferredoxin-NADP reductase.[b]
2. One-equivalent donors, four-equivalent acceptors (O_2):
    Ascorbate oxidase, laccase, ceruloplasmin, cytochrome oxidase.
3. Two-equivalent donors, one-equivalent acceptors:
    NADH dehydrogenase,[b] cytochrome b_5 reductase,[b] NADPH-cytochrome c
    reductase.[b]
4. Two-equivalent donors, two-equivalent acceptors:
    Xanthine oxidase,[c] D-amino acid oxidase, glucose oxidase, old yellow enzyme,[c]
    galactose oxidase, dihydroorotate dehydrogenase,[b] L-amino acid oxidase,[b]
    lactate oxidase,[b] NAD(P)H dehydrogenase.[d]
5. Two-equivalent donors, four-equivalent acceptors:
    Tyrosinase.

This table involves the cases in which a conclusive identification of the electron transfer mechanism has not yet been made.   [a] In general, it is not appropriate to include dehydrogenases in this table because of their reversibility.   [b] Enzymes newly added to Mason's table.   [c] These enzymes are not pertinent enough to be included in this category for the reason discussed in the text.   [d] When the quinone is assumed to be a physiological electron acceptor.

the intermediate cases such as xanthine oxidase and old yellow enzyme; thirdly, it cannot characterize all the features of reactions of a flavoprotein such as ferredoxin-NADP reductase and dihydroorotate dehydrogenase.

In the following categories, one may be able to summarize the data obtained experimentally on the free-radical nature of primary products derived from donors or acceptors during an enzyme reaction. The one-electron mechanism means a reaction in which a molecule with an even number of electrons undergoes a one-electron oxidation or reduction under the indicated conditions and is only converted to its free radical form as a primary product. The free radical thus formed is freed of the enzyme. In general, this can be judged by its ESR spectrum and also by its oxidation-reduction activity. Of course, the term "one-electron mechanism" will be applied to the reaction in which a one-electron donor is oxidized or a one-electron acceptor is reduced.

The two-electron mechanism, on the other hand, means that a molecule with an even number of electrons undergoes a two-electron oxidation or reduction during an enzyme reaction under the indicated

TABLE III

One-Electron and Two-Electron Mechanism in the Enzyme-Catalyzed Electron Transfer Processes

1. One-electron mechanism ($\kappa = 2$)

A. Oxidation of electron donor

Electron donor to be oxidized	Oxidizing system	Method	Reference
Ascorbate	peroxidase and H_2O_2	ESR	Yamazaki and Piette (*16*)
Indoleacetate	peroxidase and H_2O_2	Fe^{3+} reduction	Yamazaki and Souzu (*18*)
NADH	peroxidase and H_2O_2	Fe^{3+} reduction	Ohnishi *et al.* (*20*)
Hydroquinone	peroxidase and H_2O_2	cytochrome *c* reduction	Ohnishi *et al.* (*20*)
Ascorbate	ascorbate oxidase and O_2	ESR	Yamazaki and Piette (*16*)
Ascorbate	ascorbate oxidase and O_2	cytochrome *c* reduction	Yamazaki (*19*)
Hydroquinone	laccase and O_2	ESR	Nakamura (*32*)

B. Reduction of electron acceptor

Electron acceptor to be reduced	Reducing system	Method	Reference
p-Benzoquinone	NADH cytochrome b_5 reductase and NADH	ESR	Iyanagi and Yamazaki (*68*)
p-Benzoquinone	NADH cytochrome b_5 reductase and NADH	cytochrome *c* reduction	Iyanagi and Yamazaki (*22*)
p-Benzoquinone	NADPH cytochrome *c* reductase and NADPH	ESR	Iyanagi and Yamazaki (*68*)
p-Benzoquinone	NADH dehydrogenase and NADH	ESR and cytochrome *c* reduction	Iyanagi and Yamazaki (*22*)
p-Benzoquinone	ferredoxin-NADP reductase and NADPH	cytochrome *c* reduction	Iyanagi and Yamazaki (*22*)

2. Two-electron mechanism ($\kappa = 0$)[a]

A. Oxidation of electron donor

Electron donor to be oxidized	Oxidizing system	Method	Reference
Catechol	O_2 and tyrosinase	ESR	Mason *et al.* (*37*)

B. Reduction of electron acceptor

Electron acceptor to be reduced	Reducing system	Method	Reference
p-Benzoquinone	NAD(P)H dehydrogenase and NADH	ESR and cytochrome c reduction	Iyanagi and Yamazaki (*22*)
2-Methyl 1, 4-naphthoquinone		O_2 consumption	
2-Methyl 1, 4-naphthoquinone	NAD(P)H dehydrogenase and NADPH	O_2 consumption	Nishibayashi *et al.* (*53*)
O_2	D-amino acid oxidase and D-amino acid	cytochrome c reduction	Massey *et al.* (*65*)
	L-amino acid oxidase and L-amino acid		
	glucose oxidase and glucose		
	glycollate oxidase and glycallate		
H_2O_2	peroxidase and various reducing agents	——b)	

3. *Mixed mechanism (κ lies between 0 and 2)*[c)]
Reduction of electron acceptor

Electron acceptor to be reduced	Reducing system	Method	Reference
O_2	xanthine oxidase and xanthine[d)]	cytochrome c reduction	Handler *et al.* (*56*) McCord and Fridovich (*57*) Massey *et al.* (*65*)
p-Benzoquinone		peroxidase Compound III formation	Nakamura and Yamazaki (*21*)
		ESR and cytochrome c reduction	
O_2	aldehyde oxidase and aldehyde	cytochrome c reduction	Handler *et al.* (*56*)
O_2	dihydroorotic dehydrogenase and NADH	cytochrome c reduction	Miller and Massey (*63*)
O_2	old yellow enzyme and NADPH	cytochrome c reduction	Massey *et al.* (*65*)

a) The mechanism of O_2 reduction in the reactions of ascorbate oxidase, laccase, tyrosinase and cytochrome oxidase will be classified as a four-electron mechanism. This is concluded from the results that H_2O_2 does not occur in an appreciable amount during the reaction and also is not consumed in the reaction. b) H_2O_2 appears to be reduced in a two-step process at the surface of the peroxidase molecule but the intermediate is not detached from the enzyme. c) Deviations from each extreme case ($\kappa = 0$ or 2) are beyond experimental errors. d) As can be seen in Fig. 15, the κ for the reduction of p-benzoquinone reaches almost 2 when the concentration of p-benzoquinone is 100 μM while it is close to zero in the presence of a few μM of p-benzoquinone.

conditions and its primary product that is released from the enzyme system is a valence-saturated species.

A mixed mechanism implies that the reaction of a specific molecule proceeds in both mechanisms at the same time. Practically, this is the case where the value of κ deviates from 0 and 2 beyond experimental errors. The κ is not necessarily constant for the specific enzyme reaction as was clearly shown in the reaction of p-benzoquinone and O_2 with the xanthine oxidase system. Therefore, it becomes particularly important to indicate the experimental conditions in which the κ is measured.

It is obvious that the present classification is directly based on experimental data that can be obtained by up-to-date methods and characterizes quantitatively the electron transfer processes in enzymic oxidation-reduction reactions under the specified conditions. It can be safely said that individual reaction will be characterized in our manner, while the enzyme is characterized in Mason's manner. The present classification seems to be an elementary one that underlies the Mason-type classification in electron transferring enzymes.

When O_2 is reduced in biochemical systems, the two- and four-electron problem becomes very important in both electron transferring enzymes (66) and oxygenases (38). According to the present classification of electron transferring enzymes, both mechanisms belong to the " $\kappa=0$ mechanism." Identification of the four-electron mechanism is generally not so difficult, as was described in the section on copper enzymes, and has been made in the case of several oxidases as has been summarized by LuValle and Goddard (67) and Mason (66). Special care is required in confirming the two-electron mechanism that has been discussed in this report. The details of the present classification are listed in Table III.

DISCUSSION

The distinction between one-electron and two-electron mechanisms made in this report is a tentative one that is closely related to the experimental data. The two-electron mechanism does not imply " simultaneous " transfer of two electrons. The distinction is clearly explained in Fig. 2 on the basis of the intrinsic behavior of an imaginary intermediate which involves the free radical species in question. This intermediate can be either a binary complex or a ternary complex. In the latter case, a concerted reaction between donor and acceptor might be expected at the surface of the enzyme. At any rate, by the one-electron mechanism is meant the reaction in which the free-radical species is released into the medium.

The problem will then arise as to whether there is a sharp distinction between free radicals bound with the enzyme and free in the solution. In general, it may be said that the problem has not yet been solved. If the discussion is limited to p-benzosemiquinone, there is much evidence that the semiquinones under consideration are free in the solution. The reasons are as follow: (1) They show clear ESR hyperfine structure. This assumption is based on a criterion for the frequency of the rotational relaxation of free radical molecules. Ehrenberg (69) has suggested that the rotational relaxation time for the EMN radical in an aqueous solution can be estimated to be on the order of 10^{-9} sec. This implies that in this case anisotropic terms of up to about 30 gauss will be averaged out effectively, while only anisotropic terms of the order of a few tenths of a gauss can be averaged out for the FMN radicals of the old yellow enzyme. The relaxation time of the protein was considered to be on the order of 10^{-7} sec. Ehrenberg has observed a great difference in ESR signal between free and enzyme-bound FMN free radicals. (2) In the reactions of peroxidase and cytochrome b_5 reductase, the steady state concentration of p-benzosemiquinone exceeds the enzyme concentration by about ten times. This result excludes the possibility that the p-benzosemiquinone observed is bound with the enzyme at its active site. (3) The dismutation constant of the p-benzosemiquinone measured during the enzyme reaction is extremely consistent with that obtained in the enzyme-free system (17). (4) The rate constant of the reaction of the p-benzosemiquinone with cytochrome c has been found to be about 3×10^6 M^{-1} sec^{-1} using different methods in enzymic and nonenzymic reactions.

From these observations, it might be concluded that the p-benzosemiquinone molecule we observed is uniform and exists freely in an aqueous solution. On the contrary, there is a possibility that such free radicals might occur in various forms at the surface of the enzyme. It seems pertinent, however, to discuss this problem after such enzyme-bound free radicals are identified experimentally.

Regardless of the nature of an electron transfer complex of an enzyme, it seemed reasonable to generalize the reaction of oxidation or reduction of a molecule in the schematic diagram in Fig. 2. The problem now is what are the possible factors by which the fate of imaginary enzyme-bound free radicals is decided. The following three factors will be discussed as a temporary measure that decides whether the free radical is detached from the enzyme before or after transferring one more electron.

First, when the active sites that undergo a one-electron oxidation-reduction cycle are located separately on the enzyme surface, the one-electron mechanism will become probable for the reaction of a molecule that utilizes the cycle. Does this explain the formation of free radicals of a donor in the reactions of ascorbate oxidase, laccase and celuroplasmin? Since these enzymes contain more than 4 cupric ions per molecule that are reduced upon the addition of each donor at a reasonable rate, the stoichiometric formation of free radicals from the donor appears to be explained by this idea. However, then how can it be explained that the primary product of O_2 reduction is H_2O? In connection with this problem, an interesting observation has been reported by Nakamura and Ogura (70) to the effect that the copper that is responsible for the oxidase activity is characterized by an absorption band at 330 mμ in the difference spectrum. It might be suggested from their kinetic data that the donor transfers an electron to blue copper but O_2 reacts with another type of copper. If so, the problem is how electrons move from the blue copper to the " oxidase copper " during catalytic reaction.

Second, it seems to be a necessary prerequisite for the one-electron mechanism that the enzyme or the enzyme system take a stable intermediate after giving or accepting a single electron. Peroxidase kinetics have been studied extensively by Chance (10) who has shown that the reaction of Compound I with a donor is much faster than the reaction of Compound II with a donor. This implies that the observable intermediate in the steady state of peroxidase reaction is Compound II but not Compound I. This compound appears to be the only one that has been identified as a kinetically compulsory intermediate related with the one-electron mechanism. Much has been discussed, regarding the kind of intermediate that might appear in the reactions of flavoproteins. Upon the addition of physiological electron donors, most of the metal flavoproteins give ESR signals due to their flavin and their non-heme iron (60, 61, 71–75). Though these signals are taken mostly in frozen solution, it seems very likely that the flavin free radicals will appear during their catalytic reaction. Because of experimental difficulties, however, little has been reported on the appearance of such flavin free radicals in the steady state of the reaction. It should be emphasized here that there is little correlation between the metal and the one-electron mechanism.

In connection with the present discussion, it may be permissible to cite from interesting discussions made by Beinert and Massey in the " Flavins and Flavoproteins " symposium (76). Contrary to the old concept that the metal in the flavoproteins is needed to accomplish one-electron transfer to one-electron acceptors, Beinert has suggested the opposite, namely that the metal in the enzymes is needed, or at least is useful, in a two-electron transfer. He also suggested that a two-electron transfer would be facilitated by the flavin-iron-sulfur complex of the iron-flavoproteins ; the flavin and the iron-sulfur complex each accepting one electron in rapid succession or, for practical purposes, simultaneously. Massey agreed with Beinert, stating that one of the generalizations would be that while the initial reduction of flavoproteins by substrates is a two-electron reduction, still in the catalytic reactions in which they are involved the flavin, appears in most cases to be involved at the semi-quinone level. It has been also suggested by Massey that the state is achieved in a variety of ways, depending on the degree of sophistication of the enzyme system: (1) by two flavin semiquinones functioning at one catalytic site as in L-amino acid oxidase; (2) by flavin semiquinone-substrate complexes as in D-amino acid oxidase; (3) by flavin semi-quinone-mercaptide complexes as in lipoamide dehydrogenase and gluta-thione reductase; (4) by flavin semiquinone-metal-mercaptide inter-actions in the case of the metal flavoproteins.

In general, it may be considered that these complexes are apt to form a fairly stable intermediate after it transfers a single electron to the acceptor. This idea seems to be applicable to most cases of flavopro-teins that catalyze a one-electron transfer to the acceptor. A new spec-tral species observed by Strittmatter (47) might be such an intermediate. A similar case has been reported by Masters et $al.$ (41), who stated that NADPH-cytochrome c reductase contains 2 moles of FAD per mole of the enzyme and that the overall reaction involves the operation of a $2FADH \leftrightarrow 2FADH_2$ " shuttle " for ferricyanide as well as cytochrome c reduction. It was also found by Kamin et $al.$ (77) and by Masters et $al.$ (41) that menadione utilizes the same $2FADH \leftrightarrow$

$2FADH_2$ shuttle for its reduction as do the one-electron acceptors. The one-electron mechanism for reduction of p-benzoquinone by this enzyme will be based on the stability of an intermediate, FADH·FADH$_2$. Experimental confirmation of such an intermediate, however, has not yet been made.

In connection with the above problems, it is pertinent to ask whether or not the stability of free radicals derived from a donor or an acceptor modifies the way in which the electron transfer process in enzyme reactions is carried out. When free radicals exist free in an aqueous solution, their stability will be expressed in terms of its semiquinone formation constant (4) or its dismutation rate constant. Free radicals of indole-acetate and pyridine nucleotides appear to be less stable than p-benzo-semiquinone at the surface of an enzyme as well as in aqueous solution. But it was confirmed, as listed in Table III, that indoleacetate and NADH are oxidized in peroxidase reactions according to the one-electron mechanism. It was also shown in xanthine oxidase reactions (21) that O_2 and p-benzoquinone are reduced according to similar reaction mechanisms although there is probably a considerable difference in the stability between O_2^- and p-benzosemiquinone. At the moment, it might be said that the mechanism is affected mostly by the stability of the inter-mediates that occur in the enzyme system. This hypothetical mecha-nism is incomplete in many respects and must be substantiated by further study.

Third, the most important factor that influences the electron transfer process would be the lifetime of an intermediate complex that involves a free radical of a donor or an acceptor under consideration. Much of the discussion centered around this problem (11, 13, 14, 79). An estimate of the probable lifetime of such an intermediate cannot be given by experimental methods for the time being. Weiss (14) has suggested that in the case of the two uncharged reactants the half-time cannot be less than the time of relaxation of the water dipoles in liquid water, which is on the order of 10^{-10}–10^{-11} sec, and also that the time will be a lower limit, because in the interaction of water molecules with an ion the time of relaxation in the immediate neighbor-hood of the ion will have to be considered. Kozyrev (79) has pointed out that these relaxation times can be considerably greater than those measured for the pure solvent and in certain cases they can be as long as 10^{-7} sec.

According to Weiss (*14*), the jump frequency for atoms (or ions) is given by

$$\nu = \nu_0 \exp^{-E/RT} ,$$

where E is the activation energy and ν_0 (frequency coefficient) is on the order of 10^{13}–10^{14} sec^{-1} for a typical atom (or ion) transfer. Thus, for example, even in a primary complex of the relatively short lifetime of 10^{-10} sec, about $10^{-10} \times 10^{13} = 10^3$ jumps can take place if the process goes on without appreciable energy for activation. Even up to activation energies of about 4 kcal, it would still be highly probable that an intramolecular transfer of a heavy particle would take place. Weiss (*14*) has then pointed out the probability that during the lifetime of the primary complex, an intramolecular transfer of atoms or radicals will take place without any appreciable participation of the solvent.

According to the assumption of Weiss (*14*) and the present definition, most of the oxidation-reduction reactions should occur in the two-electron mechanism. Though Weiss's assumption may not explain the results reported in this paper, there still exists a strong possibility that the lifetime of Weiss's primary complex must be one of the most important factors that affect the electron transfer mechanism under consideration. It might be very roughly concluded from Table III that if a molecule is a specific substrate for an enzyme the molecule is detached from the enzyme in some forms of valence saturation. The reduction of O_2 in O_2-obligative oxidase reactions, the reduction of H_2O_2 in peroxidase reactions and the oxidation of electron donors in most of flavoprotein reactions appear to be good examples. It can be said that such a molecule forms a specific and stable complex with an enzyme or with an enzyme system during its catalytic reaction. This will be achieved through various physical and chemical interactions at the surface of the enzyme. Under these circumstances, a primary formed one-electron transfer complex may remain in the bound state until the subsequent electron transfer occurs.

On the other hand, a great many molecules are oxidized in peroxidase reactions or are reduced in reactions of soluble flavoproteins that are isolated from electron transport systems. One may conclude, therefore, that the reactions of these molecules with the corresponding enzyme or enzyme system proceed probably without any specific interaction that might occur in the ordinary enzyme-substrate reactions. Consequently, the primary complex will have a relatively short lifetime and will release the free radical molecule into the medium. However, this hypothetical theory is quite incomplete in many respects since it cannot explain the distinction in the mechanism of quinone reduction by the following flavoproteins: NADH dehydrogenase, xanthine oxidase and NAD(P)H dehydrogenase.

Two of the most important questions are why NAD(P)H dehydrogenase catalyzes a two-electron reduction of quinones while it also catalyzes a rapid reduction of ferricyanide $(51, 52)$ and why the κ depends upon the acceptor concentrations in xanthine oxidase reactions. In connection with the latter, it was suggested by Handler $et\ al.$ (56) that there are two terminal sites for external acceptors in xanthine oxidase; but a possibility that such sites are non-heme iron was excluded from the findings of Komai $et\ al.$ (59), who suggested direct electron transfer from a reduced flavin to O_2. It might be pertinent to ask whether a mixed mechanism is explained on the basis of a single catalytic site (refer to Fig. 2 and Eqs. 3 and 4) or on the basis of a mixture of two catalytic sites (or two isozymes) that perform either a one-electron or two-electron transfer catalysis. This will be related to the similar questions of whether or not the κ value is invariable for many enzyme reactions listed in Table III, even under the extreme experimental conditions. At the moment it remains an unsolved problem.

SUMMARY

On the basis of experimental data, electron transfer reactions catalyzed by enzymes are classified into three mechanistic categories: a one-electron transfer mechanism, a two-electron transfer mechanism and a mixed mechanism. The parameter, κ, was introduced to characterize the mechanism quantitatively. When a molecule undergoes bivalent oxidation or reduction in enzyme reactions, the κ equals the ratio of the rate of free radical formation to the rate of overall enzyme reaction at the steady state. The rate of free-radical formation is measured in two different ways: first, by measuring the steady state concentration of the free radical with ESR spectroscopy and, secondly, by measuring the rate of reduction of an iron-complex which is added as a scavenger of the free radical. A number of reactions are classified in the present report but it is still too early to generalize about the mechanisms on the basis of some definite criterion. The studies on enzyme kinetics in relation to peroxidase reactions that have been done will have to be extended on the other enzyme reactions in order to gain a general conception of the mechanism of electron transfer catalysts.

ACKNOWLEDGEMENTS

The author wishes to express his gratitude to Dr. Takashi Iyanagi for the valuable discussions they shared.

REFERENCES

1 F. Haber, *Naturwiss.*, **19**, 450 (1931).
2 F. Haber and R. Willstätter, *Ber. Deut. Chem. Ges.*, **64**, 2844 (1931).
3 L. Michaelis, *J. Biol. Chem.*, **96**, 703 (1932).
4 L. Michaelis, in " The Enzymes," ed. by J. B. Sumner and K. Myrbäck, Academic Press, New York, Vol. 2, Part I, p. 1 (1951).
5 R. Kuhn and T. Wagner-Jauregg, *Ber.*, **67**, 361 (1934).
6 E. Haas, *Biochem. Z.*, **290**, 291 (1937).
7 A. Ehrenberg and G. D. Ludwig, *Science*, **127**, 1177 (1958).
8 H. Beinert and R. H. Sands, in " Free Radicals in Biological Systems," ed. by M. S. Blois *et al.*, Academic Press, New York, p. 17 (1961).
9 P. George, *Nature*, **169**, 612 (1952).
10 B. Chance, *Arch. Biochem. Biophys.*, **41**, 416 (1952).
11 B. Chance, in " The Mechanism of Enzyme Action," ed. by W. D. McElroy and B. Glass, Johns Hopkins University Press, Baltimore, p. 389 (1954).
12 I. Yamazaki, H. S. Mason and L. H. Piette, *Biochem. Biophys. Res. Commun.*, 1, 336 (1959); *J. Biol. Chem.*, **235**, 2444 (1960).
13 F. H. Westheimer, in " The Mechanism of Enzyme Action," ed. by W. D. McElroy and B. Glass, Johns Hopkins University Press, Baltimore, p. 321 (1954).
14 J. Weiss, *Nature*, **181**, 825 (1958).
15 B. Chance, in " Free Radicals in Biological Systems," ed. by M. S. Blois *et al.*, Academic Press, New York, p. 1 (1961).
16 I. Yamazaki and L. H. Piette, *Biochim. Biophys. Acta*, **50**, 62 (1961).
17 I. Yamazaki and T. Ohnishi, *Biochim. Biophys. Acta*, **112**, 469 (1966).
18 I. Yamazaki and H. Souzu, *Arch. Biochem. Biophys.*, **86**, 294 (1960).
19 I. Yamazaki, *J. Biol. Chem.*, **237**, 224 (1962).
20 T. Ohnishi, H. Yamazaki, T. Iyanagi, T. Nakamura and I. Yamazaki, *Biochim. Biophys. Acta*, **172**, 357 (1969).

21 S. Nakamura and I. Yamazaki, *Biochim. Biophys. Acta*, **189**, 29 (1969).

22 T. Iyanagi and I. Yamazaki, *Biochim. Biophys. Acta*, **216**, 282 (1970).

23 F. Haber and J. Weiss, *Proc. Roy. Soc. (London)*, **A147**, 332 (1934).

24 P. George, *Biochem. J.*, **54**, 267 (1953).

25 B. Chance, H. Theorell and A. Ehrenberg, *Arch. Biochem. Biophys.*, **37**, 237 (1942).

26 P. George, *J. Biol. Chem.*, **201**, 427 (1953).

27 T. Yonetani, H. Schleyer and A. Ehrenberg, *J. Biol. Chem.*, **241**, 3240 (1966).

28 Y. Maeda and Y. Morita, *Biochem. Biophys. Res. Commun.*, **29**, 680 (1967).

29 Y. Maeda, T. Higashimura and Y. Morita, *Ann. Repts. of Res. Reactor Inst. Kyoto Univ.*, **2**, 67 (1969).

30 T. Yonetani, H. Schleyer, A. Ehrenberg and B. Chance, in "Hemes and Hemoproteins," ed. by B. Chance, R. W. Estabrook and T. Yonetani, Academic Press, New York, p. 293 (1967).

31 T. Nakamura, *Biochem. Biophys. Res. Commun.*, **2**, 111 (1960).

32 T. Nakamura, in "Free Radicals in Biological Systems," ed. by M. S. Blois *et al.*, Academic Press, New York, p. 169 (1961).

33 L. Broman, B. G. Malmström, R. Aasa and T. Vänngård, *Biochim. Biophys. Acta*, **75**, 365 (1963).

34 Y. Ogura and T. Nakamura, in "Magnetic Resonance in Biological Systems," ed. by A. Ehrenberg, B. G. Malmström and T. Vänngård, Pergamon Press, p. 205 (1967).

35 G. Foester, W. Weiss and Hj. Staudinger, *Ann. Chem.*, **690**, 166 (1965).

36 G. R. Stark and C. R. Dawson, in "The Enzyme," ed. by P. D. Boyer, H. Lardy and K. Myrback, Academic Press, New York, **8**, 297 (1963).

37 H. S. Mason, E. Spencer and I. Yamazaki, *Biochem. Biophys. Res. Commun.*, **4**, 236 (1961).

38 G. A. Hamilton, *Advan. in Enzymology*, **32**, 55 (1969).

39 H. Beinert and G. Palmer, *Advan. in Enzymology*, **27**, 105 (1965).

40 C. H. Williams, Jr. and H. Kamin, *J. Biol. Chem.*, **237**, 587 (1962).

41 B. S. S. Masters, H. Kamin, Q. H. Gibson and C. H. Williams, Jr., *J. Biol. Chem.*, **240**, 921 (1965).

42 K. Yagi, in "Flavins and Flavoproteins," ed. by E. C. Slater, Elsevier, Amsterdam, p. 210 (1966).

43 T. E. King, *J. Biol. Chem.*, **238**, 4032 (1963).

44 T. P. Singer and E. B. Kearney, in "Methods of Biochemical Analysis," ed. by D. Glick, Interscience, New York, p. 307 (1957).

45 Y. Hatefi, K. E. Stempel and W. G. Hanstein, *J. Biol. Chem.*, **244**, 2358 (1969).

46 R. L. Pharo, L. A. Sordahl, H. Edelhoch and D. R. Sanadi, *Arch. Biochem.*

Biophys., **125**, 416 (1968).

47 P. Strittmatter, in " Flavins and Flavoproteins," ed. by E. C. Slater, Elsevier, Amsterdam, p. 325 (1966).

48 I. Yamazaki, K. Yokota and R. Nakajima, in " Oxidases and Related Redox Systems," ed. by T. E. King, H. S. Mason and M. Morrison, John Wiley, New York, p. 485 (1965).

49 K. Yokota and I. Yamazaki, *Biochim. Biophys. Acta*, **105**, 301 (1965).

50 C. Martius, in " The Enzymes," ed. by P. D. Boyer, H. Lardy and K. Myrbäck, Academic Press, New York, Vol. 7, p. 517 (1963).

51 L. Ernster, L. Danielson and M. Ljunggren, *Biochim. Biophys. Acta*, **58**, 171 (1962).

52 F. Marki and C. Martius, *Biochem. Z.*, **333**, 111 (1960).

53 H. Nishibayashi, T. Omura and R. Sato, *J. Biochem.*, **60**, 172 (1966).

54 I. Fridovich and P. Handler, *J. Biol. Chem.*, **236**, 1836 (1961).

55 L. Greenlee, I. Fridovich and P. Handler, *Biochemistry*, **1**, 779 (1962).

56 P. Handler, K. V. Rajagopalan and V. Aleman, *Federation Proc.*, **23**, 30 (1964).

57 J. M. McCord and I. Fridovich, *J. Biol. Chem.*, **243**, 5753 (1968).

58 P. F. Knowles, J. F. Gibson, F. M. Pick and R. C. Bray, *Biochem. J.*, **111**, 53 (1969).

59 H. Komai, V. Massey and G. Palmer, *J. Biol. Chem.*, **244**, 1692 (1969).

60 G. Palmer, R. C. Bray and H. Beinert, *J. Biol. Chem.*, **239**, 2657 (1964).

61 R. C. Bray, G. Palmer and H. Beinert, *J. Biol. Chem.*, **239**, 2667 (1964).

62 E. Ackerman and A. S. Brill, *Biochim. Biophys. Acta*, **56**, 397 (1962).

63 R. W. Miller and V. Massey, *J. Biol. Chem.*, **240**, 1466 (1965).

64 J. M. McCord and I. Fridovich, *Federation Proc.*, **28**, 346 (1969).

65 V. Massey, S. Strickland, S. G. Mayhew, L. G. Howell, P. C. Engel, R. G. Matthews, M. Schuman and P. A. Sullivan, *Biochem. Biophys. Res. Commun.*, **36**, 891 (1969).

66 H. S. Mason, *Ann. Rev. Biochem.*, **34**, 595 (1965).

67 J. E. LuValle and D. R. Goddard, *Quart. Rev. Biol.*, **23**, 197 (1948).

68 T. Iyanagi and I. Yamazaki, *Biochim. Biophys. Acta*, **172**, 370 (1969).

69 A. Ehrenberg, *Ark. Kemi*, **19**, No. **7**, 97 (1962).

70 T. Nakamura and Y. Ogura, *J. Biochem.*, **64**, 267 (1968).

71 T. E. King, R. L. Howard and H. S. Mason, *Biochem. Biophys. Res. Commun.*, **5**, 329 (1961).

72 H. Beinert and R. H. Sands, *Biochem. Biophys. Res. Commun.*, **3**, 41 (1960).

73 K. V. Rajagopalan, V. Aleman and P. Handler, *Biochem. Biophys. Res. Commun.*, **8**, 220 (1962).

74 V. Aleman, S. T. Smith, K. V. Rajagopalan and P. Handler, in " Flavins and Flavoproteins," ed. by E. C. Slater, Elsevier, Amsterdam, p. 99

(1966).

75 D. V. Dervartanian, W. P. Zeylemaker and C. Veeger, in "Flavins and Flavoproteins," ed. by E. C. Slater, Elsevier, Amsterdam, p. 183 (1966).

76 H. Beinert and V. Massey, in "Flavins and Flavoproteins," ed. by E. C. Slater, Elsevier, Amsterdam, p. 97 (1966).

77 H. Kamin, B. S. S. Masters and Q. H. Gibson, in "Flavins and Flavoproteins," ed. by E. C. Slater, Elsevier, Amsterdam, p. 306 (1966).

78 P. George and J. S. Griffith, in "The Enzyme," ed. by P. D. Boyer, H. Lardy and K. Myrbäck, Academic Press, New York, p. 347 (1959).

79 B. M. Kozyrev, *Discussions Faraday Soc.*, **19**, 135 (1955).

Received for publication, April 10, 1970.

Advan. in Biophys., Vol. 2, pp. 77–112 (1971)

THE ELECTROGENIC SODIUM PUMP

KYOZO KOKETSU

Department of Physiology, Kurume University School of Medicine, Kurume, Japan

1. The Concept of the Electrogenic Sodium Pump

The mechanism underlying the maintenance of the resting potential and the production of the action potential has been successfully interpreted by the ionic theory (*cf. 1, 2*), which is now widely accepted in the field of cell membrane physiology. This theory is originally based on the following basic concept, which may be very hypothetical in regard to the biophysics and biochemistry of neural tissues. Namely, according to this concept, the cytoplasm contains a high concentration of anions that are provided with various kinds of amino acids that cannot penetrate the cell membrane. On the other hand, the resting membrane is readily permeable to potassium and chloride ions and it is sparingly permeable to sodium ions. The movement of potassium and chloride ions across the resting membrane is solely driven by the electrochemical forces of these ions, distributed in the extra- and intracellular fluids. Similarly, the sodium ions in the extracellular fluid also move into the cytoplasm, according to their electrochemical forces.

The ionic theory proposes an important concept in connection with the outward movement of sodium ions through the cell membrane. Namely, a sodium pump extrudes sodium ions from the intracellular fluids as fast as they diffuse into the cytoplasm. Such a sodium pump is assumed to be operated by the biological energy provided with cellular metabolism and to maintain the intracellular sodium concentration at approximately 10% of the extracellular concentration. On the other hand, potassium and chloride ions are assumed to be distributed asymmetrically in the extracellular and intracellular spaces under the condition of Donnan equilibrium. Thus, the value E of the resting potential of the cell membrane would be equal to the potassium equilibrium potential, E_K, or to the chloride equilibrium potential, E_{Cl}, expressed as follows:

$$E = E_K = -\frac{RT}{F} \log_e \frac{(K)_i}{(K)_o} = E_{Cl} = -\frac{RT}{F} \log_e \frac{(Cl)_o}{(Cl)_i}, \qquad (1)$$

where $(K)_i$ and $(K)_o$ are the concentrations of potassium ions in the intra- and extracellular fluids, respectively. The same is true for the chloride ions. R, T and F are the gas constant, absolute temperature and Faraday's constant, respectively.

A number of investigations were carried out with various excitable cells in order to examine the applicability of Eq. 1 to the actual value observed directly from living cells in the resting state. These investigations have confirmed that the observed value of the resting potential deviates appreciably from the value of the potassium or chloride equilibrium potential.

Hodgkin and Katz (3) have shown that this deviation is to be expected because the excised tissues are no longer in a steady state, but are gaining sodium and losing potassium. Thus, they have proposed that in such an unsteady state the resting potential would be given approximately by the following equation, derived from Goldman's constant-field theory (4):

$$E = -\frac{RT}{F} \log_e \frac{P_K(K)_i + P_{Na}(Na)_i + P_{Cl}(Cl)_o}{P_K(K)_o + P_{Na}(Na)_o + P_{Cl}(Cl)_i}, \qquad (2)$$

where P_K, P_{Na} and P_{Cl} are the permeability constants for the respective ions; and $(Na)_i$ and $(Na)_o$ are the concentrations of sodium ions in the intra- and extracellular fluids, respectively.

It has been shown in many subsequent experiments, carried out with various kinds of excitable cells in the resting state, that the value predicted by Eq. 2 agrees fairly well with the observed value of the resting potential.   It must be kept in mind, however, that Eq. 2 can be derived on assuming that (1) the outward movement of sodium through the membrane driven by the presumed sodium pump is electrically neutral, that is, it produces no net transfer of charge across the membrane, and (2) the sum of ionic currents contributed by the movements of sodium, potassium and chloride ions, driven by their electrochemical forces, is zero (3).

It is important to examine the validity of the first assumption. If the outward movement of sodium is not electrically neutral, that is, if the sodium pump produces a net transfer of charge across the membrane, the observed value of the resting potential would deviate from the value predicted by Eq. 2.   Such a possibility cannot be easily discarded, since many experimental results, which suggest that the outward movement of sodium, at least a fraction of it, is not electrically neutral, have been recently accumulated.

Whether or not the sodium pump produces a net transfer of charge across the membrane would depend on how the sodium pump worked. The sodium pump could operate in one of two alternative ways.   It could extrude sodium not in ionic form but in combination with some intracellular anions or some kinds of negatively charged carriers.   Such a sodium pump would cause no outward sodium current, so the outward movement of sodium (the sodium efflux) would be electrically neutral. This kind of sodium pump may be described as electrically neutral. On the other hand, the sodium pump could extrude sodium as ions in a way that would cause an outward sodium current.   In this case, the sodium efflux is not electrically neutral, so it would be an electrogenic sodium pump.

However this sodium pump operates, the sodium efflux driven by the sodium pump appears to be coupled with the inward movement of potassium (the potassium influx) (1, 5).   Experimental evidence, which showed that the potassium influx was not simply driven by an electrochemical force, was first reported with mammalian erythrocytes (cf. 6). Shanes (7) also suggested that the potassium influx in the nerves might be driven by energy other than electrochemical energy.   A close coupling between the sodium efflux and the potassium influx was demonstrated by the fact that the sodium efflux was reduced upon removal of the external potassium (8–11).

The question arises as to exactly how the sodium efflux and the potassium influx are linked together. At one extreme, the sodium efflux could be tightly coupled in a 1:1 ratio with the potassium influx. In this case, there would be no separation of electrical charge across the membrane, that is, the sodium efflux would be electrically neutral. This kind of an electrically neutral coupling between the sodium efflux and the potassium influx may be described as "chemical" coupling. If this is true, the sodium pump would be electrically neutral.

At the other extreme, if the sodium pump produces an outward sodium current and creates a potential difference across the membrane—the inside becoming negative—which would attract the potassium ions, the coupling would be "electrical." Such a sodium pump or one that operated in an intermediate fashion with partial chemical coupling would be electrogenic.

The sodium efflux seems to be coupled with the potassium influx both chemically and electrically in various excitable tissues. Indeed, the possibility that the sodium efflux is not coupled 1 : 1 with the potassium influx and that, therefore, the sodium pump is capable of generating an electrical potential in the course of its operation has been reported by a number of investigators (12–15). Mullins and Noda (12) have suggested that the coupling ratio in frog skeletal muscles could not be higher than one potassium pumped in for every three sodiums pumped out. It was also suggested that the relative contribution of the chemical and electrical coupling to the total sodium and potassium movements might vary with experimental conditions (13–15).

2. *Membrane Potential and Electrogenic Sodium Pump*

A most convenient way to test whether or not a part of the membrane potential is produced by the electrogenic sodium pump would be to study the effects of various kinds of metabolic inhibitors on the membrane potential. Since the sodium pump should be operated by consuming the biological energy, it will be slowed down when the energetic cellular metabolism is inhibited by an appropriate metabolic inhibitor. Thus, if the sodium pump, even only a part of it, is electrogenic, the membrane potential would show a rapid change after an application of a metabolic inhibitor. In the case of squid giant axons, it was demonstrated that the resting membrane potential was decreased slowly, but no faster than might have been expected from a gradual depletion of the intracellular potassium concentration, suggesting that the sodium pump was not directly involved in the change in the resting membrane potential (*11*). Furthermore, it was clearly demonstrated that squid giant axons in which the sodium efflux and potassium influx have been completely inhibited with dinitrophenol or cyanid, nevertheless, showed little change in their resting potentials (*11*). These results appear to suggest that the concept expressed by Eq. 2 could be applied to the resting state of squid giant axon.

The action of the metabolic inhibitors of frog skeletal muscle fibers is not consistent with the results obtained from squid giant axons. Ling and Gerard (*16*) first demonstrated that the resting membrane potential of frog skeletal muscle fibers dropped markedly under the effects of metabolic inhibitors that inhibited glycolysis and oxidative metabolism. Similar results were also obtained by subsequent experiments (*17–19*).

However, muscles doubly poisoned with cyanide and iodoacetate, which should block both respiration and glycolysis, showed no striking reduction in sodium efflux (*20*). The resting membrane potential of the muscle fibers rapidly fell when dinitrophenol was applied to the external solution, and the size of the potential drop could not be explained by a depletion of intracellular potassium (*21*). It was confirmed, however, that the sodium efflux was not depressed but rather increased while the potassium influx was markedly depressed (*21*). These results indicated that the initial rapid depolarization observed in the presence of dinitrophenol, which markedly inhibited oxidative phosphorylation in frog skeletal muscle (*22*), was not due to an inhibition of an electrogenic sodium pump.

The reasons why the resting membrane of frog skeletal muscle fibers is rapidly depolarized in the presence of metabolic inhibitors and why the sodium efflux is not depressed in the same condition are unknown at present. Changes in the membrane permeability to ions do not seem to be responsible for a rapid initial depolarization (*21*), and it is possible that some unknown mechanisms are involved in the bioelectric potential and in the linkage between the ionic pump and cellular metabolism (*23*).

Another method for examining the possibility that the electrogenic sodium pump may be partially responsible for the establishment of the resting membrane potential is to study the effect of temperature changes on the resting membrane potential. As seen in Eq. 2, the value of the resting membrane potential would be proportional to the absolute temperature, provided that the relative membrane permeabilities of the sodium, potassium and chloride ions are unchanged by the change in temperature. Since the electrogenic sodium pump should be very sensitive to a change in temperature, if it exists, the resting membrane potential would rapidly change when the temperature is suddenly altered: the change in potential would not be proportional to the change in absolute temperature, even though the relative membrane permeabilities remain constant.

In the case of the squid giant axon, almost no change in the resting membrane potential was observed over the range of temperature from 0 to 20°C (*24, 25*). Hodgkin and Katz (*24*) suggested that the lack of change in potential in coordination with the temperature change could be accounted for by assuming that the relative ion permeabilities of the membrane were altered. Experiments with frog skeletal muscles were not conclusive either; a linear relationship between the resting membrane potential and the absolute temperature, expected according to Eqs. 1 or 2, may (*26, 27*) or may not (*28*) be demonstrated according to the experimental conditions.

Coombs *et al.* (*29*) first demonstrated the experimental evidence suggesting the presence of an electrogenic sodium pump in cat spinal motoneuron. In this experiment, a hyperpolarization of the membrane was observed immediately after an injection of sodium ions into the cell. They suggested that the raised potential might be attributed to a specific active extrusion of sodium ions carrying positive charges out of the cell, and that the level of activity of the sodium pump is determined by the internal concentration of sodium ions. Hodgkin and Keynes (*30*) also observed a small rise in the membrane potential with an intracellular injection of sodium ions into the squid giant axon and suggested a contribution of the electrogenic sodium pump to the membrane potential (*cf. 31*). The presence of an electrogenic sodium pump in various kinds of excitable cells has been suggested by a number of subsequent electrophysiological investigations, which will be described together with the experimental results which have been obtained in our laboratory.

EXPERIMENTAL EVIDENCE FOR AN ELECTROGENIC SODIUM PUMP

1. Resting Potential

A) Normal intracellular sodium concentration

Prologue—Many experimental results indicate that, when cells are in the resting state, the measured value of the resting potential is well interpreted according to Eq. 2. As has been described, it was difficult to demonstrate any convincing result, suggesting the presence of an electrogenic sodium pump in the electrophysiological experiments in which the effects of metabolic inhibitors or temperature change have been studied with squid giant axon or frog skeletal muscle in the resting state. It must be kept in mind, however, that the lack of any marked changes of the resting potential in these experiments is not conclusive evidence that the pump is not electrogenic, since it is possible that the fraction of the membrane potential contributed by the electrogenic sodium pump would be too small to be detected experimentally.

In the case of squid giant axon, two-thirds of the sodium efflux appears to be coupled chemically with the potassium influx; and, if the remaining third is actually an electrogenic sodium pump, the direct contribution of the pump to the membrane potential would be only 1.8 mV (*11*). Furthermore, the electrogenic fraction in a total sodium efflux may vary depending on the kind of cells; and it is possible that its direct contribution to the membrane potential would be much more significant in some particular kinds of cells. Indeed, there are a number of papers that report experimental evidence suggesting the contribution of the electrogenic sodium pump to the resting potential of certain kinds of cells.

Experimental results and discussion—In the case of lobster giant axon, the resting membrane was hyperpolarized rapidly and markedly when the temperature was raised (*32, 33*), and this kind of hyperpolarization became less prominent when a metabolic inhibitor, such as ouabain, dinitrophenol, cyanide or azide, was applied in advance to the extracellular solution (*33*). Accordingly, a suggestion that the temperature-dependent component of the lobster axon resting membrane potential resulted from a metabolically linked process, namely, from the electrogenic sodium pump, was forwarded by Senft (*33*).

The resting membrane potential of *Aplysia* neurons was also found to vary much more with temperature than was predicted by Eq. 2 (*34, 35*). In this experiment, it was pointed out that the resting membrane potential, increased by warming, exceeded significantly the potassium equilibrium potential, which was determined by measurement of the equilibrium point of the spike after-potential (*36*). The result demonstrating this fact is seen in Fig. 1. Carpenter and Alving (*36*)

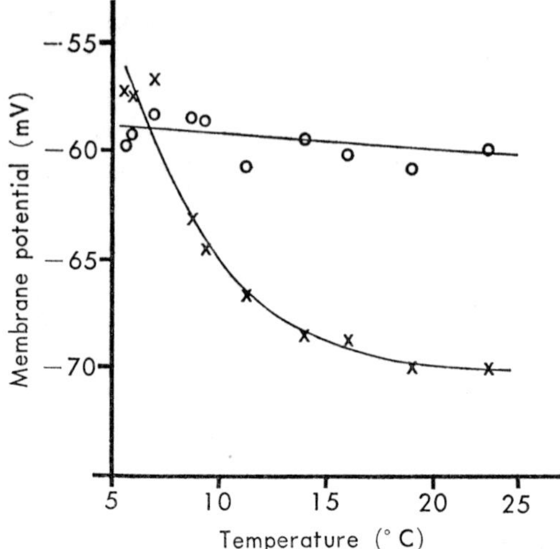

Fig. 1.   Plot of the variation of after-potential and resting membrane potential (RMP) with temperature in one experiment.   Each point is an average of three measurements (*36*).   ×, RMP;   O, after-potential.

found that such a hyperpolarization was abolished by ouabain, replacement of external sodium by lithium and removal of external potassium. Furthermore, they confirmed that the hyperpolarization was independent of the external chloride and was increased by cocaine, which increased the resistance to passive movements of potassium ions (*37*).   Thus, it was concluded that the hyperpolarization caused by warming could most reasonably be ascribed to the activity of the electrogenic sodium pump (*36, cf. 38*).   This concept has been supported by other investigators, who found similar experimental results from mulluscan neurons (*39, 40*).

Comparable results were also obtained from photoreceptor cells in *Limulus*.   It was found in this experiment that the membrane potential dependence upon temperature exceeded that attributable to Eq. 1 and that the membrane was depolarized by the action of ouabain (*41*). These data indicated that the photoreceptor cells had an electrogenic sodium pump that contributed a fraction to the total membrane potential (*41*).   In the *Limulus* photoreceptor cells, removing sodium or potassium or increasing calcium in the extracellular sodium depolarizes the membrane;   it was assumed that these procedures inactivated adenosine triphosphatase and, thereby, removed the electrogenic component of the sodium pump (*41*).

The internal potential of *Neurospora* was reported to have two components, namely, the first component which was reduced by anoxia or abolished by respiratory inhibitors, and the second component which remained in the presence of respiratory inhibitors and was sensitive to the external potassium concentration (*42*). Respiratory inhibition by dinitrophenol, azide or carbon monoxide causes the internal potential to drop more than 80% in one minute, whereas the ionic distribution changes at a rate of less than 30% per hour, *i.e.*, more than 200 times more slowly (*42–44*).

It was noted that the close coupling between oxidative metabolism and the trans-surface potential difference in *Neurospora* was reminiscent of the situation found with the endocochlear potential (*45*), the gastric potential (*46*) and postanoxic hyperpolarization of frog skin (*47*) or myelinated nerves (*48*); in all these cases electrical potentials shifted more rapidly than would be expected from gross changes of ion concentration (*42*). These results appear to suggest the presence of an electrogenic sodium pump.

The resting potential of toad spinal ganglion cells increased rapidly, approximately 20 mV, when the external sodium ions were partially or totally replaced by equimolar barium ions; such a marked hyperpolarization can be observed in a solution containing 10 mM (millimol. per liter H_2O) potassium with or without the chloride ions (*49*). The hyperpolarization was abolished at a low temperature and by the action of dinitrophenol, while it was augmented when the intracellular sodium concentration was increased by injecting sodium into a cell. Thus, it was suggested that the barium-induced hyperpolarization was due to the electrogenic sodium pump (*49*). The membranes of the barium-treated cells showed very great membrane resistance and became remarkedly insensitive to the changes in the external potassium concentration. The hyperpolarization produced by the electrogenic sodium pump was unmasked and became large enough to be recorded as a removal of the short circuit effect of passive potassium movement (*49*).

The presence of the electrogenic sodium pump in cells in the resting state has been suggested in many experiments on the basis of the experimental evidence that the membrane is hyperpolarized by warming beyond the value predicted by Eq. 2. These results, however, should be carefully reexamined, since the changes in intracellular ionic concentrations and also those in membrane permeabilities to different ions during the experiment have not yet been explored well in the tested cells. The fact that a metabolic inhibitor can prevent the hyperpolarization caused by warming or that it can directly depolarize the membrane may not also justify the presence of the electrogenic sodium pump. The changes in ionic conductances, which could be caused by the specific or non-specific actions of metabolic inhibitors, should be carefully analyzed.

B) Increased intracellular sodium concentration
Prologue—In many cases, even when the electrogenic sodium pump is present, the potential drop across the resistance of the membrane may be too small to be detected from resting muscle or nerve cells, which maintain a normal intracellular sodium concentration. The sodium pump seems to be greatly accelerated when the intracellular sodium concentration is increased either physiologically (production of the action potential) or experimentally (sodium loading or sodium injection).

There are many experimental results showing that the sodium efflux is controlled by the level of intracellular sodium. Hodgkin and Keynes (*30*) suggested that the sodium efflux was directly proportional to the intracellular sodium concentration; and Keynes and Swan (*50*) showed that over a small range of internal sodium concentration the sodium efflux was proportional to the cube of the internal sodium concentration, but suggested that over a wide range the relationship might be more directly proportional (*cf. 14, 51*). Thus, if the sodium pump could produce the potential drop across the membrane by any mechanism, such an electrical sign would be detected more easily when the internal sodium concentration was increased by any means.

Experimental results and discussion
a) Sodium-loaded cells: In frog skeletal muscle, Kernan (*52*) has shown that when the muscles are actively extruding sodium, the resting potential was significantly higher than the calculated potassium equilibrium potential. In this experiment, the membrane potentials of muscle fibers, made rich in sodium by soaking overnight in a cold potassium-free Ringer solution, were measured on reimmersion in a recovery solution containing 10 mM potassium (*cf. 53–55*). A significant difference between the observed potential and the calculated potassium

equilibrium potential was also observed in a chloride-free recovery solution where all of the chloride ions were replaced by sulfate, and a marked potential difference was observed with the sodium-free recovery solution (containing choline, chloride or sucrose) in which the loss of intracellular sodium ions was greatly facilitated (52).

It was also found that the resting potential during sodium extrusion was significantly higher than in the muscles treated with the metabolic inhibitor, o-phenanthroline (56). Thus, Kernan (52, 56) has concluded that the elevated resting potential that takes place during sodium extrusion is due to the operation of an electrogenic sodium pump and suggested that the chloride as well as potassium ions, moving passively to equilibrium at the end of the period of extrusion of sodium, do not contribute to the observed potential. The experimental fact that in frog muscle during the replacement of accumulated sodium by potassium in a recovery solution the value of the measured membrane potential is significantly more negative than the potassium equilibrium potential has been confirmed by subsequent papers (13–15, 57–59). The presence of an electrogenic sodium pump was also suggested by the sodium-loaded cat heart muscle which showed a hyperpolarization beyond the likely potassium equilibrium potential while it was extruding excess intracellular sodium (60).

In our laboratory, experiments were carried out with skeletal muscles of warm-blooded animals in order to confirm the presence of an electrogenic sodium pump. In one of these experiments, an isolated rat diaphragm preparation was soaked in a cold (5°C) potassium-free Tyrode solution for 7 hr and perfused with a cold glutamate Tyrode solution (total chloride ions in the Tyrode solution were replaced by equimolar glutamate ions) for 2 hr. The perfusion solution was thereafter changed to a cold glutamate Tyrode solution containing 10 mM potassium. A microelectrode was inserted into a surface fiber and the membrane potential was recorded continuously while the perfusion solution was switched to a warm (35°C) glutamate Tyrode solution containing 10 mM potassium. Ouabain was then added to the perfusion solution, and the microelectrode was withdrawn. An example of results obtained from these experiments is shown in Record B of Fig. 2.

The glutamate ions were used in our experiments because the condition of the muscles was better maintained using these ions in

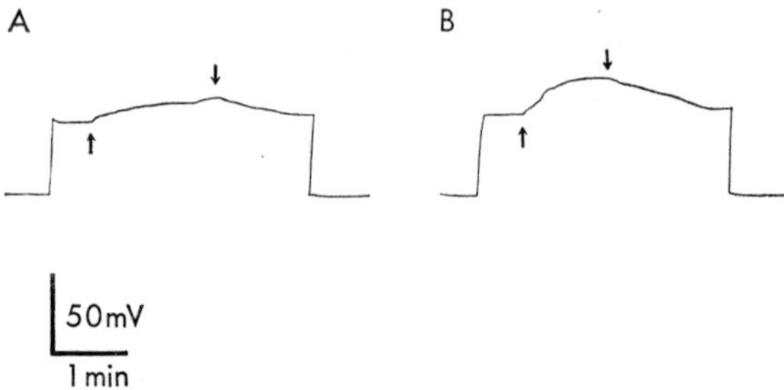

Fig. 2.   Pen-writing recording showing the hyperpolarization of the rat dia-phragm muscle membrane, produced by a sudden warming. The preparations were previously soaked in a cold potassium-free Tyrode solution for 7 hr, and the experiments were carried out in a 10 mM potassium Tyrode (A) and 10 mM potassium glutamate Tyrode solutions (B). The moments of insertion and withdrawal of a microelectrode are seen at the beginning and at the end of the records, respectively. The temperature of the perfusing solutions was changed from 5 to 35°C at the moments indicated by the upward arrows and ouabain (0.05 mM) was applied at the moments indicated by the downward arrows.

comparison with the sulfate ions which had been used in other experi-ments (*14, 15, 52*). Preliminary experiments showed that when the muscles were soaked in the glutamate Tyrode solution at room tempera-ture (32°C), the membrane was rapidly depolarized, showing a transient chloride potential (*cf. 61*). It finally exhibited a steady resting potential whose value was approximately -90 to -95 mV, within 1 hr (*cf. 62*). The effective membrane resistance measured at a steady state in the glutamate Tyrode solution was approximately 150% of that measured in a normal Tyrode solution. Thus, the membrane permeability to the glutamate ions could be assumed to be much smaller than that to the chloride ions (*cf. 63, 64*). The value of the resting potential measured in the cold 10 mM K glutamate Tyrode solution would be very close to the potassium equilibrium potential expressed by Eq. 1. The average value of the resting potential measured from these muscles was -47 mV. On the other hand, the value measured from muscles soaked in the same solution at the same temperature, without a previous soaking in the cold potassium-free solution, was -58 mV. These results indicated that the intracellular potassium concentration was approximately 80 mM and 120 mM for the muscles with and without a previous soaking in a cold potassium-free solution, respectively.

As seen in Record B of Fig. 2, the membrane is rapidly hyper-polarized when the temperature of the external solution (10 mM K glutamate Tyrode solution) is abruptly raised to 35°C.   In three experiments, the absolute value of the resting potential increased 23 mV, on the average, within one minute.   Such an increase in the resting potential is not compatible with the temperature coefficient of Eq. 1.   Indeed, the

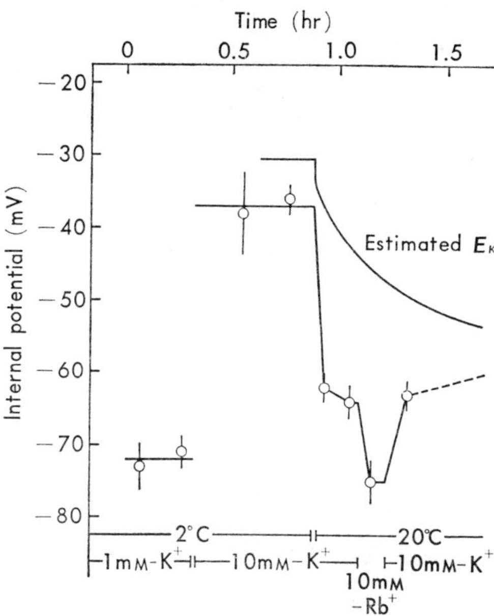

Fig. 3.   The internal potential of the surface fibers of sodium-rich, chloride-depleted muscles in sulfate solutions containing 10 m-equiv./1 of either potassium or rubidium.   The muscle had been soaked for 42.5 hr in a potassium-free Ringer solution and then for 2 hr in a chloride-free sulfate Ringer solution before any potential measurements were made.   Potassium or rubidium concentrations and the temperature of the solution are indicated at the bottom of the figure.   After 1.6 hr the experimental muscle had an internal potassium concentration of 66 mequiv./kg fiber water.   Its pair, analyzed at 0 hr, had an internal potassium concentration of 36 mequiv./kg fiber water. These values have been used to estimate the initial and final values of the potassium equilibrium potential; the time course of changes in the potassium equilibrium potential was estimated on the basis of results obtained from separate experiments.   The sodium concentrations of the control and experimental muscle were 70 and 45 mequiv./kg muscle.   Each point is the mean (±S. E. of mean) of not less than six measurements (*15*).

resting potential seems to show a more negative value than the expected potassium equilibrium potential immediately after a rise of temperature; the concentration of intracellular potassium would not increase so rapidly as to alter the potassium equilibrium potential appreciably, even in the surface fibers extruding sodium. As seen in Record B of Fig. 2, the hyperpolarization was rapidly reduced upon the addition of ouabain, suggesting that the hyperpolarization is related directly to the metabolic processes.

In order to correlate the hyperpolarization observed upon warming of the recovery solution with the intracellular potassium and sodium concentrations, the actual changes in these ionic concentrations were studied with frog skeletal muscles in many experiments (*13–15, 56–58*). According to the experimental results, the hyperpolarization apparently exceeded the value of the potassium equilibrium potential calculated from the measured intracellular potassium concentration. The data shown in Fig. 3 were reported by Adrian and Slayman (*15*); and they demonstrate that the internal potential is almost 30 mV more negative than the estimated potassium equilibrium potential, immediately after warming in a chloride-free sulfate Ringer solution containing 10 mM potassium.

Two alternative hypotheses were proposed here for the explanation of the observed hyperpolarization (*15*). The outward sodium current caused by the electrogenic sodium pump may be responsible for the membrane hyperpolarization. Alternatively, the electrically neutral sodium pump, which causes the sodium efflux coupled chemically with the potassium influx, removes the extracellular potassium immediately outside of the surface fibers and thus increases the potassium equilibrium potential and consequently the resting potential. Such a view was first proposed for the interpretation of posttetanic hyperpolarization in mammalian C fibers (*65*). When the sodium pump is accelerated, the rate of potassium movement is consequently larger. Thus, the interpretation of the observed hyperpolarization is always complicated by the effects of extracellular diffusion. The situation will be more complex if the pumping takes place in the walls of the transverse tubules of the sarcoplasmic reticulum. It is certainly very difficult to distinguish the two alternative hypotheses (*cf. 15*).

In order to assess the alternatives for the interpretation of the hyperpolarization observed with muscle fibers, experiments were undertaken on muscles exchanging rubidium, rather than potassium, for internal sodium (*15*). It was found that the maximum membrane potential observed within 20–30 min in a recovery solution containing either potassium or rubidium ranged from -75 to -90 mV in 10 mM potassium and from -90 to -105 mV in 10 mM rubidium. On the other hand, normal muscles tested in the sulfate solution had internal potentials less negative in rubidium than in potassium at equal concentrations below about 4 mM.

As seen in Fig. 3, the substitution of rubidium for potassium in the sulfate recovery solution produced additional hyperpolarization that is reversible. Since the rubidium moved into cells nearly as fast as the potassium and since the permeability of muscle membrane to rubidium was much less than its permeability to potassium, the greater hyperpolarization in fibers exchanging rubidium for sodium than in fibers exchanging potassium for sodium was qualitatively consistent with an electrogenic sodium pump and the greater membrane resistance to the movement of rubidium (*15*). Thus, it was concluded that an electrically neutral sodium pump cannot be wholly responsible for the hyperpolarization observed during pumping (*15*).

The effects of change in the external potassium concentration on the sodium-induced hyperpolarization do not provide any conclusive evidence for separating the two hypotheses. The removal of the external potassium has been shown to inhibit the sodium pump to a certain extent in frog skeletal muscle (*10, 50*), suggesting that the electrically neutral sodium pump is inhibited. It would be expected that, in a potassium-free solution, further depletion of potassium in the vicinity immediately outside of the membrane by the operation of the electrically neutral sodium pump is precluded. Thus, if the electrically neutral sodium pump is responsible for the sodium-induced hyperpolarization, such a hyperpolarization would be diminished in a potassium-free solution; the membrane potential should not exceed the resting potential observed in this solution. On the other hand, if the electrogenic sodium pump is responsible for the sodium-induced hyperpolarization, the removal of the external potassium would cause little change in the hyperpolarization.

In our experiments, no further hyperpolarization was observed when the temperature was raised in the glutamate Tyrode solution containing no potassium, as has been reported in the experiments with frog skeletal muscle (*15*).   The absence of the sodium-induced hyperpolarization in the potassium-free solution is very suggestive.   This is, however, not conclusive enough to validate the possibility that the sodium-induced hyperpolarization is caused by the electrically neutral sodium pump for the following reasons.   It is possible that an electrogenic sodium pump may be activated in combination with the electrically neutral sodium pump.   It is also possible that an electrogenic sodium pump may be activated only when the potassium ions are present in the external solution; the potassium ions may react to the membrane surface where the same kind of carrier is located.   If either one of these two pos-sibilities was the case, the sodium-induced hyperpolarization produced by the electrogenic sodium pump would also be inhibited in a potassium-free solution (*15*).

The electrogenic sodium pump, if it exists, constitutes a current which contributes directly to the membrane potential by an amount which will depend upon the membrane resistance (*11*).   Since the muscle membrane permeability to the chloride ions is very high in a solution containing the chloride ions, the sodium-induced hyperpolarization produced by any mechanism would cause a passive outward movement of chloride in order to equalize the electrochemical potential of the chloride inside and outside the fiber.   The hyperpolarization, therefore, will be less than it would be if there were no chloride ions present (*15*). In our experiments, when they were carried out in the 10 mM K Tyrode solution (containing chloride), the sodium-induced hyperpolarization was observed to develop more slowly and its amplitude was smaller than the hyperpolarization observed in a chloride-free glutamate solution (see A of Fig. 2).   Although this result is consistent with the hypothesis that the hyperpolarization is produced by the electrogenic sodium pump (*cf. 77*), it does not necessarily support this hypothesis since the local potassium gradient produced by an electrically neutral sodium pump will also be less than it would be if there were no chloride ions present (*cf. 15*).

If the sodium-induced hyperpolarization is produced by the elec-trogenic sodium pump, there should be an inward potassium movement. Such a potassium movement would reduce the hyperpolarization pro-duced by the electrogenic sodium pump; such a short-circuiting effect of the potassium movement on the hyperpolarization would be depend-ent on the potassium conductance of the muscle membrane.   Thus, if the electrogenic sodium pump is responsible for the hyperpolarization, the hyperpolarization should be increased and maintained for a longer period when the potassium conductance is decreased by any means. If, on the other hand, the electrically neutral sodium pump is responsible for the hyperpolarization, there should be no enhancement of the hyperpolarization even when the potassium conductance is decreased. If the potassium conductance could be decreased by some treatment which does not affect the sodium pump mechanism, such a treatment would be very useful for judging whether the electrogenic or electrically neutral sodium pump is responsible for the hyperpolarization.

Cocaine reduces the potassium conductance of the muscle membrane and makes it relatively insensitive to changes in membrane potential (*37, 67*). Thus, cocaine should, if its only action is on potassium conductance, increase the amount of hyperpolarization produced by an electrogenic sodium pump. Adrian and Slayman (*15*) confirmed this expectation; an addition of 0.1% cocaine increased the hyperpolarization 15–20 mV. They found that, if the potassium equilibrium potential is taken into consideration, the amount of hyperpolarization and also the amount of membrane resistance were multiplied by a factor of 2–3 over a time of 10 min in the presence of cocaine.

Harris and Ochs (*59*) have also studied the effects of local anaesthetic and antihistaminic drugs, which would decrease the potassium conductance of muscle, on the sodium-induced hyperpolarization and the membrane resistance in a chloride-free solution. They found that these drugs took some time, about 30 min, to exert their full effect, and noted since the quantitative correlation between the exact time course of the potential and resistance changes was not made, the possibility that much of the potassium movement occurred before the final high resistance had been reached. Furthermore, it is uncertain whether or not the cocaine had effected the sodium pump itself (*15*).

b) Sodium-injected cells: The intracellular sodium concentration, to which the sodium pump is proportional, can be increased by injecting the sodium ions into a cell. If the electrogenic sodium pump exists and is also proportional to the intracellular sodium concentration, it would be possible to detect the hyperpolarization produced by the pump by intracellular injection of the sodium ions (*29, 30*).

Kerkut and Thomas (*68*) studied the effects of an intracellular sodium injection on the membrane potential of snail neurons. They found that the membrane potential was markedly increased (about 30 mV) when the sodium ions were injected at the rate of 4.4 mM/min into a neuron. No such hyperpolarization was brought about when the potassium ions were injected instead; and it was also inhibited by ouabain or parachlormercuribenzoate. A similar hyperpolarization caused by a sodium injection was obtained by Nakajima and Takahashi (*69*) in the stretch receptor neurons of the crayfish and also by Pinsker and Kandel (*39*) in *Aplysia* neurons.

Recently, Thomas (*70*) has confirmed the earlier observation on snail neurons. In his experiment, the membrane current during the hyperpolarization produced by a sodium injection was measured together with the intracellular sodium activity, estimated by use of a sodium-sensitive glass microelectrode. It was concluded that about two-thirds of the sodium pump was chemically coupled with the transport of other ions, probably with the influx of potassium, the remaining third producing the electrogenic effect.

A marked hyperpolarization obtained when sodium was injected into a snail neuron is shown in Fig. 4 (*70*). The membrane potential

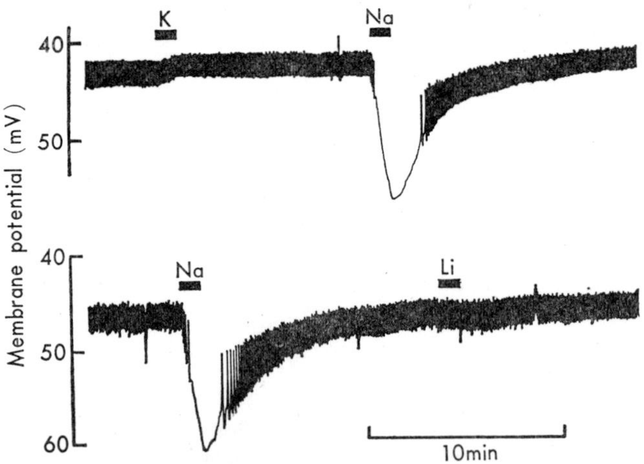

Fig. 4. Pen recording of the membrane potential of snail neurons from two experiments, showing the effects of injections of potassium acetate, sodium acetate and lithium acetate. The long time constant of the pen recorder has reduced the spontaneous action potentials to only a few mV. Solid bars indicate periods of injection; the injection current in the top half of the figure is 38 nA, and in the bottom half, 34 nA (*70*).

was recorded in this experiment with an intracellular microelectrode while the current for injecting ions was passed through two other intracellular microelectrodes.   In the case of the top record shown in Fig. 4, one injection electrode was filled with sodium acetate and the other with potassium acetate.   Thus, the passage of the current in one direction causes the injection of the sodium ions from the first electrode and acetate ions from the second, while the current passed in the opposite direction causes the injection of potassium and acetate ions.   In the case of the bottom record shown in Fig. 4, the injection electrodes were filled with sodium acetate and lithium acetate.   As seen in this figure, no hyperpolarization was observed when potassium or lithium was injected instead of sodium.

The hyperpolarization produced by a sodium injection into a snail neuron was abolished when the external potassium was removed (*68, 70*). Furthermore, a removal of external potassium did not itself cause any increase in the potential of cells.   It was suggested that these experimental results support the concept that the hyperpolarization was the result of the operation of an electrogenic sodium pump, rather than being due to a depletion of the potassium ions immediately outside the cell membrane as a result of its uptake into the cell by an electrically neutral sodium pump (*68, 70*).   However, as has been already discussed in relation to the results obtained from skeletal muscle, it is difficult to draw any conclusion from the potassium removal experiments.

2.   *Action Potential*

A)   After-hyperpolarization

Prologue—The action potential of various excitable cells is usually composed of the initial rapid spike potential and the following slow afterpotential.   The after-hyperpolarization (thereafter AH) (hyperpolarizing after-potential) has been explained as having been produced as a result of a large delayed increase in the potassium permeability during the action potential.   According to the ionic theory, the peak of the after-hyperpolarization roughly corresponds to the potassium equilibrium potential expressed by Eq. 1.   Thus, although there is a wide variation in the time course of AH following the spike potentials of different cells, the AH has generally been interpreted as being produced by a selective increase in the potassium permeability.   On the other hand, the timing of AH suggests that the AH reflects metabolic processes associated with recovery, namely, it reflects active extrusion of the sodium that enters into a cell during the spike potential. Thus, the possibility that the AH may be produced by the electrogenic sodium pump is worth taking into account in the interpretation of the mechanism underlying the AH.

Experimental results and discussion—Temporal summation of the AH's of two or three successive action potentials has been observed in a mammalian A fiber (*71*), in a cat spinal motoneuron (*72*), in a pyramidal tract cell (*73*) and crustacean stretch receptor neurons (*69*). These results suggested that the AH of action potentials in these preparations is composed of a component which is responsible for building up the posttetanic hyperpolarization (PTH), which is probably associated with sodium extrusion (see Sec. 2-B).

Greengard and Straub (*74*) recorded the monophasic action potentials of mammalian sympathetic C fibers using the sucrose-gap method and found that the spike potential was followed by a negative after-potential (after-depolarization) and also a prolonged AH whose duration was several hundred milliseconds. Such a marked AH was reversibly abolished when the external sodium ions were replaced by the lithium ions. They suggested that the AH was caused by the increased activity of the sodium pump following a single spike potential. Thus, according to the idea that was proposed for the explanation of the posttetanic hyperpolarization (*65*), the AH was explained as a result of an increase in the electrically neutral sodium pump, which depleted the potassium ions immediately outside of the axonal membrane and, thereby, created a large potassium equilibrium potential.

In a potassium-free solution, the spike potential of C fibers is followed by a brief period of positivity preceding the negative after-potential (*74*). It was suggested that this positivity is identical with the positive after-potential of the squid giant axon caused by an increase in the membrane permeability to potassium. Indeed, the positive after-potential of the squid giant axon is enhanced in a potassium-free solution and depressed in a high potassium solution (*3, 75*). A train of impulses reduces the positivity in sympathetic C fibers (*74*), in a way similar to that observed with squid giant axon (*76*).

In our laboratory, two different components of the AH in the action potentials of bullfrog sympathetic ganglion cells have been analyzed by means of an intracellular microelectrode (*77*). The action potential of

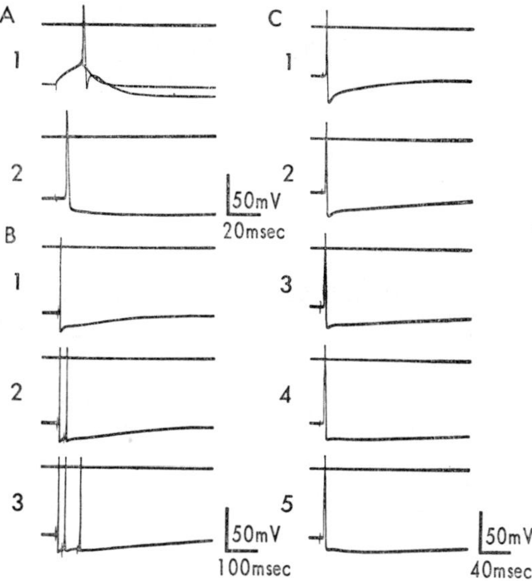

Fig. 5. After-hyperpolarization (AH) of the action potentials recorded from bullfrog sympathetic ganglion cells; note the AH which is composed of AH_1 and AH_2. A: Action potentials produced by intracellular direct (1) and antidromic indirect (2) stimulations in two different cells. In Record 1, AH_1 (initial sharp peak) and AH_2 (following blunt peak) are clearly separated. In Record 2, the peak of AH_2 is seen to be more negative (inside) than that of AH_1. B: Summation of AH_2 of action potentials produced by two (2) and three (3) successive indirect stimulations. Note the notch which separates AH_1 and AH_2 in Record 1. C: Effects of changes in the resting membrane potential level on AH_1 and AH_2. Record 3 was taken at the normal resting membrane potential, and the membrane was previously depolarized (Records 2 and 1) or hyperpolarized (Records 4 and 5) progressively (77).

these cells is composed of a spike potential which is followed by a markedly prolonged AH. The AH reaches its peak immediately after the initiation of a spike potential and returns slowly to the original resting potential level. A close inspection of the time course of the AH reveals that there is a notch which appears soon after the peak of an AH (Fig. 5). Our experimental results suggested that this notch is a sign indicating the fact that the AH consists of two different components, namely, AH_1 and AH_2.

The AH_1 is a component which forms the initial peak of AH after the initiation of the spike potential and its behavior indicates that it is a common type of AH, being produced by a selective increase of the membrane permeability to the potassium ions. Indeed, the peak level of AH_1 is fixed at a certain potential level, *viz.*, an equilibrium potential, when the resting membrane potential is shifted to different levels by passing anodal or cathodal currents (see Fig. 5).

AH_2 seems to develop more slowly than AH_1. Thus, it develops while AH_1 is diminishing, by forming a notch on the falling phase of AH_1. When the resting membrane is slightly hyperpolarized by an anodal current, the AH_2, particularly its duration, is markedly augmented. When the AH_2 is augmented by a conditioning hyperpolarization, its delayed peak level is disclosed and it often exceeds the peak level of AH_1 (Fig. 5). When two or more action potentials are produced successively by repetitive stimulations, the duration of AH_2 is significantly increased by a summation, while the peak potential of AH_1 is decreased (Fig. 5). This apparently indicates that the AH_2 is responsible for building up the PTH (see Sec. 2-B).

Our experimental analysis confirmed that the physiological and pharmacological natures of AH_2 are, indeed, essentially similar to those of the PTH of the bullfrog sympathetic ganglion cells. The characteristic nature of AH_2 is represented by the following two experimental findings. (1) The AH_2, unlike the AH_1, does not exhibit the presence of the reversal potential (equilibrium potential), namely, it increases markedly by a slight conditioning hyperpolarization while it disappears by a conditioning depolarization. (2) The AH_2, unlike the AH_1, disappears when the external sodium ions are replaced by the lithium ions; and it is sensitive to metabolic inhibitors. On the basis of these experimental findings we suggest that the underlying mechanism of the AH_2 is the same as that of the PTH and is probably generated by the electrogenic sodium pump. It should be mentioned here that the AH_2 was not significantly altered when the external potassium concentration was reduced or completely removed.

In general, it appears that the AH is constituted of at least two different components; namely, one is produced by a selective increase in the membrane permeability to potassium and the other is produced as the result of an activation of the sodium pump. There seems to be a large variation in the contribution of the second component to the AH, according to the different kinds of excitable cells. Probably, the second component is very small, for example, in the case of squid axon, whereas it is dominant, for example, in the case of bullfrog sympathetic ganglion cells. It is difficult to determine whether an electrogenic sodium pump or an electrically neutral sodium pump is responsible for the second component on the basis of the experimental results that are available at present.

B) *Posttetanic hyperpolarization*

Prologue—A train of action potentials produced by repetitive stimulation is often followed by a markedly prolonged hyperpolarization of the membrane in many excitable cells. Such a hyperpolarization following tetanic responses has been described as the positive after-potential following tetanus (*71*) in mammalian medullated fibers or the posttetanic hyperpolarization, PTH (*65*), in mammalian non-medullated fibers.

Gage and Hubbard (*78*) investigated the PTH of motor nerve terminals in the isolated rat diaphragm-phrenic nerve preparation. In this experiment, the size and time course of the PTH of nerve terminals were estimated by the changes in the threshold measured by stimulating pulses applied through a microelectrode located at the end-plate region. They found that the PTH was markedly increased and prolonged for over 20 min after removal of the external potassium ions; but, after a 35 min immersion in the potassium-free solution, there was a marked reduction of the PTH.

When the external potassium was removed in this experiment, the potassium equilibrium potential would have increased greatly with little increase in the resting membrane potential. If an increase in potassium conductance is sustained after tetanus, the membrane would be hyperpolarized toward this elevated potassium equilibrium potential. If the intracellular potassium concentration is reduced after prolonged application of a potassium-free solution, the hyperpolarization caused by an increase in potassium conductance would be reduced because of a decrease in the potassium equilibrium potential. On the basis of these assumptions, Gage and Hubbard (*78*) have concluded that the observed PTH is generated as the result of a sustained increase in the membrane permeability to the potassium ions. On the other hand, a great deal of experimental evidence has been accumulated to support the concept that the PTH of other kinds of cells may be generated by an activation of the sodium pump, presumably, the electrogenic sodium pump.

Experimental results and discussion—Ritchie and Straub (*65*) found that the PTH of mammalian non-medullated fibers was abolished in the potassium-free solution and decreased by increasing the external potassium concentrations. Furthermore, they found that the PTH was abolished by metabolic inhibitors (*cf. 79, 80*), such as dinitrophenol, azide, cyanide, iodoacetate and ouabain, and also abolished when external sodium ions are replaced by lithium ions (*cf.* also *81, 82*). Ritchie and Straub (*65*) have suggested, on the basis of their experimental results, that the PTH was produced by an electrically neutral sodium pump, which reabsorbed and depleted the potassium immediately outside the membrane and, thereby, created a large potassium equilibrium potential (*cf. 83*). Indeed, the membrane potential during the PTH never exceeded the resting potential obtained in the potassium free solution (*65*).

On the other hand, the PTH observed in vertebrate medullated nerve fibers was suggested to have been produced by the electrogenic sodium pump (*81, 82, 84*). Straub (*84*) showed that the PTH was regularly found in the potassium-free solution and could greatly exceed the resting potential observed in this solution. He also showed that the size of the PTH was directly proportional to the membrane resistance when the external potassium concentration was changed in the range from 2.5 to 11 mM. The reduction of the amplitude of PTH by increasing the external potassium concentration is compatible with the idea that the electrogenic sodium pump can be short-circuited by potassium (*84, cf. 66*).

In mammalian non-medullated nerve fibers, Holmes (*79*) found that the PTH was somewhat increased in the potassium-free solution, a fact that is contrary to the results obtained by Ritchie and Straub (*65*).

Rang and Ritchie (*66*) have recently reexamined the effects of external potassium on the PTH and have reported that the PTH often increased in the potassium-free solution but it declined quickly, leaving a small hyperpolarization. Nicholls and Baylor (*85*) have reported on the PTH, which is presumably produced by the electrogenic sodium pump, after the activity of neurons in the leech central nervous system. They noted in this report that the membrane depolarization caused by increasing the external potassium concentration was enhanced during the development of the PTH. This could be also explained by the concept that the PTH would be depressed when the electrogenic sodium pump is short-circuited by potassium.

Rang and Ritchie (*66*) have shown that the PTH is markedly increased when the external chloride is replaced by sulfate or isothionate and explained the increase in the PTH caused by removing the chloride ions as the result of eliminating the short-circuiting effect of the internal chloride ions. Furthermore, they found that potassium, applied to the nerve a few minutes after stimulation in a potassium-free solution, caused the hyperpolarization of the membrane rather than the normal depolarization. On the basis of these experimental observations, Rang and Ritchie (*66*) have concluded that the hyperpolarization is actually produced by the electrogenic sodium pump, which is dependent on the external presence of potassium and could be reduced by the short-circuiting effect of the chloride ions.

Nakajima and Takahashi (*69*) have concluded that the PTH in the stretch receptor neurons of the crayfish is produced by the electrogenic sodium pump. In this experiment, two microelectrodes were inserted in a cell, and used separately for recording potential and passing current and the effect of the shift of the membrane potential level on the PTH was studied. They found that the PTH was depressed by the replacement of sodium with lithium, by an addition of dinitrophenol, or by a reduction of external potassium, and that no changes in membrane conductance were observed during the development of PTH. An important fact that they found in this experiment was that the PTH was enhanced by conditioning hyperpolarization that showed no reversal potential. Furthermore, no change in the potassium equilibrium potential, estimated from the peak of AH following a spike potential, was detected during PTH. Nakajima and Takahashi (*69*) have stressed that these findings present strong experimental evidence to support the concept that the PTH is produced by the electrogenic sodium pump.

The effect of polarization currents on the PTH of mammalian non-medullated nerves was studied using the sucrose-gap method (*66*). When the membrane potential was increased or decreased by externally applied currents in this experiment, there was relatively little change in the PTH; the amplitude of PTH was slightly reduced during conditioning hyperpolarization and also during conditioning depolarization.

In our experiments, the PTH of the bullfrog sympathetic nerves and ganglion cells has been studied (*86*). The membrane potential changes in the sympathetic nerve fibers and ganglion cells have been recorded

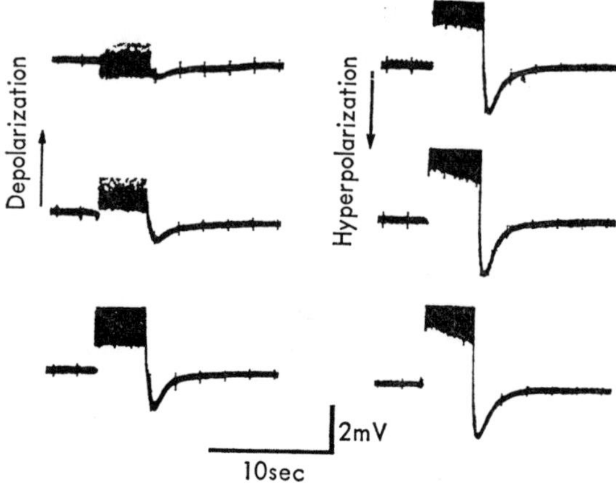

Fig. 6. Posttetanic hyperpolarization (PTH) recorded from bullfrog pre-
ganglionic sympathetic B nerve fibers by use of the sucrose-gap method.
Nerve stimulation was applied at the frequency of 30/sec for approximately
5 sec. Note the changes in peak amplitude of PTH during conditioning anodal
and cathodal currents. See text for more details (86).

using the sucrose-gap (87, 88) and the intracellular microelectrode
methods, respectively.

In the case of sympathetic nerve fibers, the PTH was recorded
from either pre- or post-ganglionic B nerve fibers (89). The PTH
recorded from these sympathetic B fibers was found to be very insensitive
to the external potassium ions. Thus, no appreciable changes in the
PTH were observed when the external potassium concentration was
reduced to 0.2 mM (one-tenth normal) or when it was raised to 20 mM.
This result was quite different from that obtained from the PTH of
other preparations. The PTH of the present preparation, however, was
abolished when the external sodium ions were replaced by the lithium
ions and depressed by the addition of metabolic inhibitors to the external
solution. The unique nature of the PTH was demonstrated when the
resting membrane potential level was shifted to a different level by
applying either anodal or cathodal conditioning current (Fig. 6). As
seen in Fig. 6, the amplitude of PTH is markedly increased when the
resting membrane is hyperpolarized by a conditioning anodal current,
while it is depressed and finally abolished when the resting membrane
is depolarized by a conditioning cathodal current.

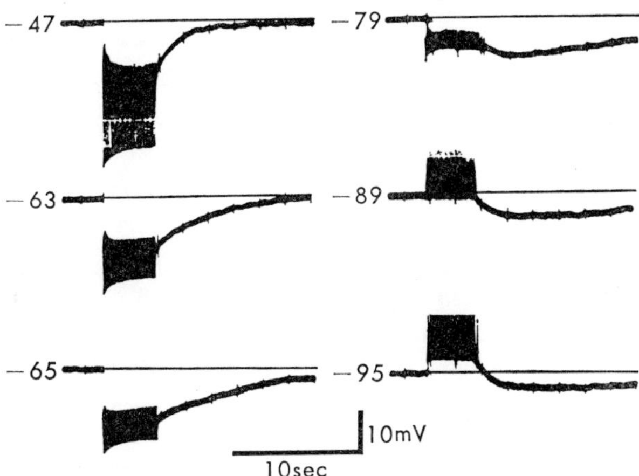

Fig. 7.   Posttetanic hyperpolarization (PTH) recorded intracellularly from a
bullfrog sympathetic ganglion cell.   Antidromic nerve stimulation was applied
at the frequency of 30/sec for approximately 5 sec.   Membrane potential was
previously depolarized or hyperpolarized by the conditioning cathodal or
anodal currents, respectively; the resting membrane potential levels (in mV)
are shown in each record.   See text for more details (86).

The experimental results, which are essentially similar to those
obtained from sympathetic B fibers, were obtained from the intracellular
analysis of the PTH generated from sympathetic ganglion cells.   The
nature of the PTH recorded from ganglion cells was also very similar to
that of the AH_2 recorded from the same cells.   Thus, the PTH of
ganglion cells was unaffected by the change in the external potassium
concentration and was depressed by replacing the external sodium ions
with the lithium ions or by the application of metabolic inhibitors.   No
significant change in the membrane conductance was observed during
the development of the PTH.   The PTH, particularly its duration, was
markedly increased when the resting membrane was hyperpolarized by a
conditioning anodal current, while it was decreased when the membrane
was depolarized by a conditioning cathodal current (Fig. 7).   These
results, together with those obtained from the sympathetic B fibers, are
consistent with the concept that the PTH of these preparations may
be produced by an activation of the electrogenic sodium pump.

3. *Postsynaptic Potential*

Prologue—There are two different kinds of chemical synaptic transmissions, namely, excitatory and inhibitory synaptic transmissions. In both cases, the chemical transmitter released from the presynaptic knobs acts on the subsynaptic membrane producing postsynaptic potentials. According to the concept based on the ionic theory, both excitatory and inhibitory postsynaptic potentials (EPSP and IPSP) are produced as a result of the increase in the membrane permeability of the subsynaptic membrane to one or more ion species which then moves their electrochemical gradients without requiring metabolic energy (*cf. 90*). This idea has been well supported by experimental evidence showing the changes in ionic conductances, which cause a shift of the membrane potential toward a certain equilibrium potential during postsynaptic potential. Indeed, numerous types of postsynaptic potentials recorded from various kinds of excitable cells have been successfully explained along the lines of this concept (*cf. 90*).

An entirely different concept of chemical synaptic transmissions has been proposed recently for explaining the mechanism underlying the production of slow IPSP, which is recorded from sympathetic ganglion cells (*88, 91, 92*). Namely, it was suggested that the slow IPSP of bullfrog sympathetic ganglion cells may be produced as a result of the activation of the electrogenic sodium pump in cells. Presumably, the chemical transmitter which acts on the subsynaptic membrane accelerates the electrogenic sodium pump, producing a transient hyperpolarization (IPSP) of the membrane.

In fact, the possibility that a chemical transmitter may alter cellular metabolism and, thereby, produce a postsynaptic potential has been considered as a way of explaining the mode of action of a transmitter. Adrenalin causes a hyperpolarization of the heart muscles (*93*); and experimental evidence has been presented suggesting that this is due to an effect on the sodium pump (*94–96*). Graham and Lamb (*97*) showed that ouabain caused a decrease in the membrane potential of frog heart muscles and prevented the hyperpolarization caused by the action of adrenalin. They suggested that these results may be explained by supposing that the activity of the sodium pump made some direct contribution to the membrane potential either by keeping the extracellular potassium concentration near the membrane low or by electrogenic action. Adrenalin also caused a hyperpolarization of smooth muscle, and it has been suggested that this hyperpolarization may be due to an acceleration of the active ion transport mechanism involving the electrogenic sodium pump (*98*, see also *99*).

It should be mentioned here in connection with the membrane potential of smooth muscles that slow fluctuation potential changes that can be recorded from the longitudinal muscle layer of the small intestine may be dependent on the electrogenic sodium pump (*100–102*). It should also be mentioned that the possibility that the end-plate potentials of frog and rat skeletal muscles may be generated by the activity of an electrogenic potassium pump activated by acetylcholine has been speculated (*103*, and see also *104, 105*).

Experimental results and discussion—The slow IPSP of bullfrog sympathetic ganglion cells has been analyzed in our experiments by observing the P potential, which is the slow IPSP recorded extracellularly by means of a sucrose-gap method (*88, 91, 92*). It was difficult to record the slow IPSP intracellularly by a microelectrode inserted into the cell-body of these cells (*cf. 88, 106*). The P potential was diminished and eventually abolished by the conditioning depolarization of ganglion cells, whereas it was enhanced by moderate conditioning hyperpolarization and was diminished and eventually abolished by intense conditioning hyperpolarization (see Fig. 8). It was also found in this experiment that in both Ringer and potassium-rich Ringer solution the abolition of the P potential by conditioning hyperpolarization occurred at a level beyond the potassium equilibrium potential, as indicated by the reversal potential of the AH_1 of the ganglionic action potential (Figs. 8 and 9).

Similar results were also obtained from preparations immersed in the chloride-free solution, in which no change in the P potential was observed. These results suggested that the P potential might be generated by an activation of the electrogenic sodium pump, being triggered by the action of the chemical synaptic transmitter. It was also found that the P potential was sensitive to ouabain and temperature changes (*92*), and it was diminished when the external potassium concentration was reduced (*88*). These results, however, do not provide any strong evidence to support the idea that the P potential is associated with the sodium pump, since the changes in the release of the transmitter have not yet been clarified under these experimental conditions.

It is interesting that the augmentation of potential change, being

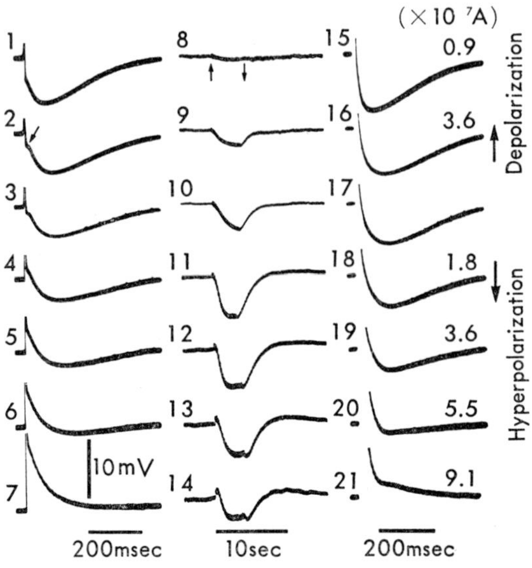

Fig. 8. Changes in the P potential and after-hyperpolarization of a ganglion by conditioning depolarization and hyperpolarization; these records were taken using the sucrose-gap method. Left column (Records 1–7): After-hyperpolarization of orthodromic responses (arrow in Record 3 indicates hump formed by EPSP). Middle column (Records 8–14): P potentials. Right column (Records 15–21): After-hyperpolarization or directly induced ganglionic responses. Spikes of orthodromic and directly induced responses are out of the frame. Orthodromic responses were induced by single stimuli applied to pre-ganglionic B fibers in the Ringer solution. P potentials and direct responses were recorded approximately 40 min after nicotine sulfate (0.12 mM) was added to the external solution. P potentials were elicited by a train of repetitive stimulation (10/sec for 4 sec) of pre-ganglionic B plus C fibers (upward and downward arrows in Record 8 indicate initiation and cessation of pre-ganglionic stimulation). Direct responses were induced by supramaximal stimulating currents (10 msec in duration) delivered through the input bridge-devise. Records in the third horizontal row, 3, 10, and 17, were taken without applying external currents, whereas those above and below this row were obtained during application of cathodal and anodal currents, respectively. The intensity of the applied currents is the same for records in the same row and is indicated in the right-hand record (*88*).

observed with the slow IPSP produced during the moderate conditioning hyperpolarization, is also seen in the cases of the AH_2 and PTH. Similarly, the absence of the equilibrium potential (reversal potential)

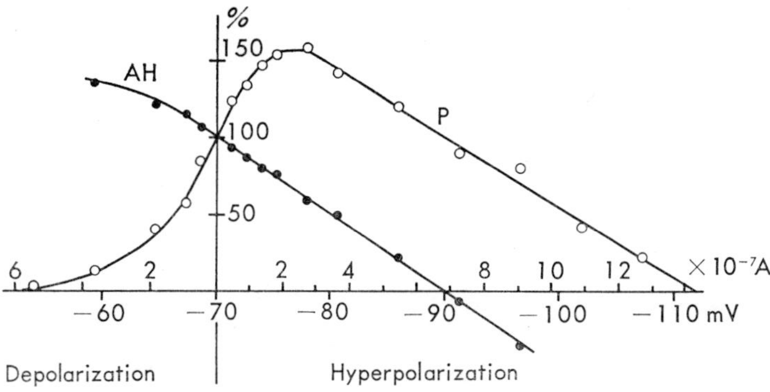

Fig. 9.   Relationship between P potential amplitude and the estimated membrane potentials of ganglion cells.   P potential (open circles) and after-hyperpolarization (AH) (solid circles) are expressed as percentages of control amplitudes (ordinate) and are plotted against the intensity of applied currents (abscissa).   This result was constructed from the experimental data that were partially illustrated in Fig. 8 (*88*).

was also observed in the cases of AH_2 and PTH.   Thus, the underlying mechanism of the slow IPSP appears to be essentially similar to that of AH_2 or PTH, which seems to be produced by the electrogenic sodium pump.   Further extensive investigation, however, should be undertaken in order to clarify if changes in the membrane conductances could be responsible for these potentials.

The idea that the slow IPSP of sympathetic ganglion cells is produced by an activation of the electrogenic sodium pump (*88, 91, 92*) has been challenged by Kobayashi and Libet (*107*), who studied the slow IPSP recorded from the sympathetic ganglia of bullfrogs and rabbits. Kobayashi and Libet (*107*) have reported that ouabain and other metabolic inhibitors, such as dinitrophenol, sodium, azide and anozia, show no specific depressive effects on the slow IPSP, and also that the application of a K-free solution does not selectively depress the slow IPSP of bullfrog sympathetic ganglion cells.   On the basis of these experimental findings, they concluded that the possibility that the electrogenic sodium pump is generating the slow IPSP can be ruled out.   They have analyzed the slow IPSP that was recorded from rabbit sympathetic ganglia and confirmed that the slow IPSP was decreased in amplitude by progressive depolarization.   It was brought to the zero level with about 20 mM depolarization from the usual resting membrane potential of about 45 mV. On the other hand, the slow IPSP was increased by a moderate hyperpolarization of 10–20 mV above the resting potential; further hyperpolarization, however, consistently produced a decrease in the slow IPSP. They noted that the actual postsynaptic mechanisms that generated the slow IPSP still had to be uncovered since the slow IPSP was produced without any detectable changes in the membrane conductance (*107*).

A prolonged IPSP (late IPSP), which was recorded by an intracellular electrode from an interneuron of *Aplysia*, was recently reported to exhibit the characteristic nature (*50*). This late IPSP is mediated monosynaptically and generated by a chemical transmitter, acetylcholine. The late IPSP does not have an ionic equilibrium potential and does not accompany ionic conductance changes (*39*). Furthermore, they found that the late IPSP was selectively inhibited by ouabain and cooling, and was selectively blocked by prolonged washing with a K-free bathing solution. They concluded, on the basis of these experimental results, that the late IPSP is generated by an augmentation of the electrogenic sodium pump.

SUMMARY

The concept of the electrogenic sodium pump which is based on the ionic theory was introduced and discussed. The mechanism of the sodium pump, particularly that of the electrogenic sodium pump, is not entirely clear at present. Nevertheless, it was stressed that the electrogenic sodium pump, if it exists, would generate a measurable potential difference across a membrane particularly when it is augmented. A number of experimental results, which have been accumulated in many laboratories supporting the presence of the electrogenic sodium pump, were introduced together with those obtained in our laboratory.

The most convincing results suggesting the presence of the electrogenic sodium pump were those obtained from frog skeletal muscles. In these experiments, the hyperpolarization of the membrane of a sodium-loaded muscle, which is produced by warming, was extensively analyzed; and it was confirmed that such a hyperpolarization actually exceeds the potassium equilibrium potential, calculated from the estimated value of the intracellular potassium concentration. It was noted, however, that the evidence obtained in these experiments is not conclusive enough to determine whether the observed hyperpolarization is due to an activation of the electrogenic sodium pump or due to that of an electrically neutral sodium pump.

The experimental results which were obtained from various cells for the analysis of (1) the hyperpolarization produced by an intracellular sodium injection, (2) the AH, (3) the PTH and (4) the IPSP were presented and discussed. These results were consistent with the concept that these potentials might be generated by an augmentation of the electrogenic sodium pump. It must be kept in mind, however, that these results could also be interpreted as the results of an augmentation of an electrically neutral sodium pump, or even as the results of the changes in the membrane permeability to different species of ions.

The concept that the sodium pump, particularly the electrogenic sodium pump, is responsible for the changes in the membrane potential is very attractive. Further extensive investigations, however, must be carried out in order to verify this concept.

REFERENCES

1 A. L. Hodgkin, *Biol. Rev.*, **26**, 339 (1951).
2 A. L. Hodgkin, *Proc. Roy. Soc. B.*, **148**, 1 (1958).
3 A. L. Hodgkin and B. Katz, *J. Physiol.*, **108**, 37 (1949).
4 D. E. Goldman, *J. Gen. Physiol.*, **27**, 37 (1943).
5 R. D. Keynes, *J. Physiol.*, **114**, 119 (1951).
6 E. Ponder, *J. Gen. Physiol.*, **33**, 745 (1950).
7 A. M. Shanes, *J. Gen. Physiol.*, **34**, 795 (1951).
8 E. J. Harris and M. Maizels, *J. Physiol.*, **113**, 506 (1951).
9 I. M. Glynn, *J. Physiol.*, **126**, 35 (1954).
10 R. D. Keynes, *Proc. Roy. Soc. B.*, **142**, 359 (1954).
11 A. L. Hodgkin and R. D. Keynes, *J. Physiol.*, **128**, 28 (1955).
12 E. J. Mullins and K. Noda, *J. Gen. Physiol.*, **47**, 117 (1963).
13 S. R. Cross, R. D. Keynes and R. Rybová, *J. Physiol.*, **181**, 865 (1965).
14 L. J. Mullins and M. Z. Awad, *J. Gen. Physiol.*, **48**, 761 (1965).
15 R. H. Adrian and C. L. Slayman, *J. Physiol.*, **184**, 970 (1966).
16 G. Ling and R. W. Gerard, *J. Cellular Comp. Physiol.*, **34**, 389 (1949).
17 W. R. Loewenstein and A. Szent-Györgyi, *J. Cellular Comp. Physiol.*, **46**, 345 (1955).
18 W. V. MacFarlane and J. D. Meares, *J. Physiol.*, **142**, 78 (1958).
19 M. Sato, K. Sakie and T. Wada, *Japan. J. Physiol.*, **15**, 561 (1962).
20 R. D. Keynes and G. W. Maisel, *Proc. Roy. Soc. B.*, **142**, 383 (1950).

21 K. Koketsu, H. Kimizuka and R. Kitamura, *J. Cellular Comp. Physiol.*, **63**, 165 (1964).

22 L. G. Abood, K. Koketsu and K. Noda, *Am. J. Physiol.*, **200**, 431 (1961).

23 K. Koketsu, *Perspectives. Biol. Med.*, **9**, 54 (1965).

24 A. L. Hodgkin and B. Katz, *J. Physiol.*, **109**, 240 (1949).

25 R. Guttman, *J. Gen. Physiol.*, **49**, 1007 (1966).

26 G. N. Ling and J. W. Woodbury, *J. Cellular Comp. Physiol.*, **34**, 407 (1949).

27 H. P. Jenerick and R. W. Gerard, *J. Cellular Comp. Physiol.*, **42**, 79 (1953).

28 J. A. Apter and K. Koketsu, *J. Cellular Comp. Physiol.*, **56**, 123 (1960).

29 J. C. Coombs, J. C. Eccles and P. Fatt, *J. Physiol.*, **130**, 291 (1955).

30 A. L. Hodgkin and R. D. Keynes, *J. Physiol.*, **131**, 592 (1956).

31 H. Grundfest, C. Y. Kao and M. Altamirano, *J. Gen. Physiol.*, **38**, 245 (1954).

32 J. C. Dalton and D. E. Hendrix, *Am. J. Physiol.*, **202**, 491 (1962).

33 J. P. Senft, *J. Gen. Physiol.*, **50**, 1835 (1967).

34 R. W. Murry, *Comp. Biochem. Physiol.*, **18**, 291 (1964),

35 D. O. Carpenter, *J. Gen. Physiol.*, **50**, 1469 (1967).

36 D. O. Carpenter and B. O. Alving, *J. Gen. Physiol.*, **52**, 1 (1968).

37 A. M. Shanes, W. H. Freygang, Jr., H. Grundfest and F. Amatniek, *J. Gen. Physiol.*, **42**, 793 (1959).

38 B. O. Alving, *J. Gen. Physiol.*, **54**, 512 (1969).

39 H. Pinsker and E. R. Kandel, *Science*, **163**, 931 (1969).

40 M. F. Marmor and A. L. F. Gorman, *Physiologist*, **12**, 293 (1969).

41 T. G. Smith, W. K. Stell, J. E. Brown, J. A. Freeman and G. C. Murray, *Science*, **162**, 456 (1968).

42 C. L. Slayman, *J. Gen. Physiol.*, **49**, 93 (1966).

43 G. Lester and O. Hechter, *Proc. Natl. Acad. Sci.*, **45**, 1792 (1959).

44 C. W. Slayman and E. L. Tatum, *Biochim. Biophys. Acta*, **102**, 149 (1965).

45 E. G. Wever, M. Lawrence, R. W. Hemphill and C. B. Straut, *Am. J. Physiol.*, **159**, 199 (1949).

46 W. S. Rehm, in " The Cellular Functions of Membrane Transport," ed. by J. F. Hoffman, Prentice-Hall, Englewood Cliffs, New Jersey, p. 231 (1964).

47 E. J. Lund, *J. Exptl. Zool.*, **51**, 291 (1928).

48 R. Lorente de No, in "A Study of Nerve Physiology," Studies from the Rockefeller Institute for Medical Research, **131**, 1 (1947).

49 S. Nishi and H. Soeda, *Nature*, **204**, 761 (1964).

50 R. D. Keynes and R. C. Swan, *J. Physiol.*, **147**, 591 (1959).

51 R. D. Mullins and A. S. Frumento, *J. Gen. Physiol.*, **46**, 629 (1963).

52 R. P. Kernan, *Nature*, **193**, 986 (1962a).

53 J. E. Desmedt, *J. Physiol.*, **121**, 191 (1953).

54 M. J. Carey, E. J. Conway and R. P. Kernan, *J. Physiol.*, **148**, 51 (1959).
55 E. J. Conway, R. P. Kernan and J. A. Zadunaisky, *J. Physiol.*, **155**, 263 (1961).
56 R. P. Kernan, *J. Physiol.*, **162**, 129 (1962b).
57 Y. Hashimoto, *Kumamoto Med. J.*, **18**, 23 (1964).
58 A. S. Frumento, *Science*, **147**, 144s (1965).
59 E. J. Harris and S. Ochs, *J. Physiol.*, **187**, 5 (1966).
60 E. Page and S. R. Storm, *J. Gen. Physiol.*, **48**, 957 (1965).
61 A. L. Hodgkin and P. Horowicz, *J. Physiol.*, **148**, 127 (1959).
62 A. Takeuchi and N. Takeuchi, *J. Physiol.*, **154**, 52 (1960).
63 O. F. Hutter and S. M. Padsha, *J. Physiol.*, **146**, 117 (1959).
64 O. F. Hutter and D. Noble, *J. Physiol.*, **151**, 89 (1960).
65 J. M. Ritchie and R. W. Straub, *J. Physiol.*, **136**, 80 (1957).
66 H. P. Rang and J. M. Ritchie, *J. Physiol.*, **196**, 183 (1968).
67 R. H. Adrian and W. H. Freygang, *J. Physiol.*, **163**, 61 (1962).
68 G. A. Kerkut and R. C. Thomas, *Comp. Biochem. Physiol.*, **14**, 167 (1965).
69 S. Nakajima and K. Takahashi, *J. Physiol.*, **187**, 105 (1966).
70 R. C. Thomas, *J. Physiol.*, **201**, 495 (1969).
71 H. S. Gasser and H. Grundfest, *Am. J. Physiol.*, **117**, 113 (1936).
72 M. Ito and T. Oshima, *Nature*, **195**, 910 (1962).
73 K. Takahashi, *J. Neurophysiol.*, **28**, 908 (1965).
74 P. Greengard and R. W. Straub, *J. Physiol.*, **144**, 442 (1958).
75 A. L. Hodgkin and R. D. Keynes, *J. Physiol.*, **128**, 61 (1955b).
76 B. Frankenhaüser and A. L. Hodgkin, *J. Physiol.*, **131**, 341 (1956).
77 K. Koketsu and S. Minota, to be published.
78 P. W. Gage and J. I. Hubbard, *Nature*, **203**, 653 (1964).
79 O. Holmes, *Archs. Int. Physiol.*, **70**, 211 (1962).
80 P. Greengard and R. W. Straub, *J. Physiol.*, **161**, 414 (1962).
81 C. M. Connelly, *Rev. Mod. Phys.*, **31**, 475 (1959).
82 C. M. Connelly, *Proc. XII Int. Congr. Physiol. Lectures and Symposia*, **1**, 600 (1962).
83 J. M. Ritchie, in " Biophysics of Pharmacological Actions," Am. Assoc. for the Advancement of Science, Washington, D. C., p. 165 (1961).
84 R. W. Straub, *J. Physiol.*, **159**, 19 (1961).
85 J. Nicholls and D. Baylor, *Science*, **162**, 279 (1968).
86 K. Koketsu, S. Nishi and S. Minota, to be published.
87 K. Koketsu and S. Nishi, *J. Physiol.*, **96**, 293 (1968).
88 S. Nishi and K. Koketsu, *J. Neurophysiol.*, **31**, 717 (1968).
89 S. Nishi, H. Soeda and K. Koketsu, *J. Cellular Comp. Physiol.*, **66**, 19 (1965).

90 J. C. Eccles, in "The Physiology of Synapses," Academic Press, New York (1964).
91 K. Koketsu and S. Nishi, *Life Sciences*, **6**, 1827 (1967).
92 S. Nishi and K. Koketsu, *Life Sciences*, **6**, 2049 (1967).
93 J. Dudel and W. Trautwein, *Experientia*, **12**, 396 (1957).
94 M. Otsuka, *Pflügers Arch. Ges. Physiol.*, **266**, 512 (1958).
95 W. Trautwein and R. F. Schmidt, *Pflügers Arch. Ges. Physiol.*, **271**, 715 (1960).
96 H. C. Haas and W. Trautwein, *Nature*, **197**, 80 (1963).
97 J. A. Graham and J. F. Lamb, *J. Physiol.*, **197**, 479 (1968).
98 G. Burnstock, *J. Physiol.*, **143**, 183 (1958).
99 E. Bülbring and H. Kuriyama, *J. Physiol.*, **166**, 59 (1963).
100 E. E. Daniel, *Can. J. Physiol. Pharmacol.*, **43**, 551 (1965).
101 D. D. Job, *Am. J. Physiol.*, **217**, 1534 (1969).
102 J. Liu, C. L. Prosser and D. D. Job, *Am. J. Physiol.*, **217**, 1542 (1969).
103 R. P. Kernan, *Nature*, **214**, 725 (1967).
104 R. P. Kernan, *Nature*, **210**, 537 (1966).
105 M. Dockry, R. P. Kernan and S. Tangney, *J. Physiol.*, **186**, 187 (1966).
106 K. Koketsu, *Federation Proc.*, **28**, 101 (1969).
107 H. Kobayashi and B. Libet, *Proc. Natl. Acad. Sci.*, **60**, 1304 (1968).

Received for publication, May 8, 1970.

Advan. in Biophys., Vol. 2, pp. 113–153 (1971)

STRUCTURE OF TROPOMYOSIN AND ITS CRYSTAL

TATSUO OOI and SUGIE FUJIME-HIGASHI

*Institute for Chemical Research, Kyoto University, Uji, Kyoto,
Japan and the Institute of Molecular Biology, Faculty of Science,
Nagoya University, Nagoya, Japan*

At the time when a component of muscle proteins was discovered by
Bailey (*1*), the protein was named tropomyosin because it seemed to be
a prototype of myosin (*2*), which plays an important role in muscle
contraction.   This protein was extracted with a concentrated KCl solution from dried muscles after washing minced muscle with organic
solvents.   Characteristic properties of the protein are to exhibit polymerization-depolymerization phenomena depending on a salt concentration (the usual feature of fibrous proteins), the stability toward pH
and temperature, and the similarity of the amino acid composition to
that of myosin.   Furthermore, the protein can be crystallized into three
dimensional form (*2*), though any crystal of other muscle proteins has
not been grown.   At the present time, tropomyosin has been established
as one of the major components of muscle proteins together with myosin
A, actin, troponin and actinin.

As for the functional aspect of tropomyosin, it has no enzymic
property.   It was thought that it played a role in keeping the structure
of myofibrils or was involved in the catch mechanism for the case of

paramyosin, a family of tropomyosin (*3*), until recently native tropomyosin, another muscle protein which had similar properties to tropomyosin, was shown to have an effect on superprecipitation of the actomyosin system by Ebashi (*4*).   Presumably, because of a lack of function normally associated with a protein such as enzymic activity, not many investigations have been done on this protein except for studies on the molecular characteristics and some mechanisms of polymerization. However, it does play an important role in the control of muscle contraction through the interactions with actin and troponin, the latter being a component protein of native tropomyosin found by Ebashi *et al.* (*5, 6*).   Thus future studies are expected to be developed on the structure and function of this protein in association with its interaction with troponin and actin.

A family of tropomyosin, extracted from muscles other than those of vertebrates, is designated as paramyosin, or tropomyosin A, whereas tropomyosin from vertebrates is often called tropomyosin B (*7*). Paramyosin has properties similar to those of tropomyosin but differs in that its solubility is restricted in a narrow region against a salt solution.

In this article, we will mainly discuss tropomyosin from rabbit skeletal muscle on which many investigations have been done as a typical example and which may be a suitable one for the study of the structure of tropomyosin.

ISOLATION AND PURIFICATION

The original method described by Bailey (*2*) for isolating tropomyosin was the extraction of the protein with a concentrated KCl (1 M) solution from dried muscle obtained by treatment of minced muscle with ethanol and ether. The extraction could be done either from acetone dried muscle after removal of actin (*8*) or from native tropomyosin (*6*).   The use of a concentrated salt solution for extraction seems to be quite reasonable because tropomyosin molecules are dispersed in depolymerized form at a high concentration of salts.   In view of its depolymerization at a high temperature, extraction can be done by elevating the temperature; for example, metin, a protein similar to native tropomyosin, can be obtained from the extract at a high temperature (*9*).

The purification process consists essentially of the repetition of two processes: isoelectric precipitation and salting out with ammonium

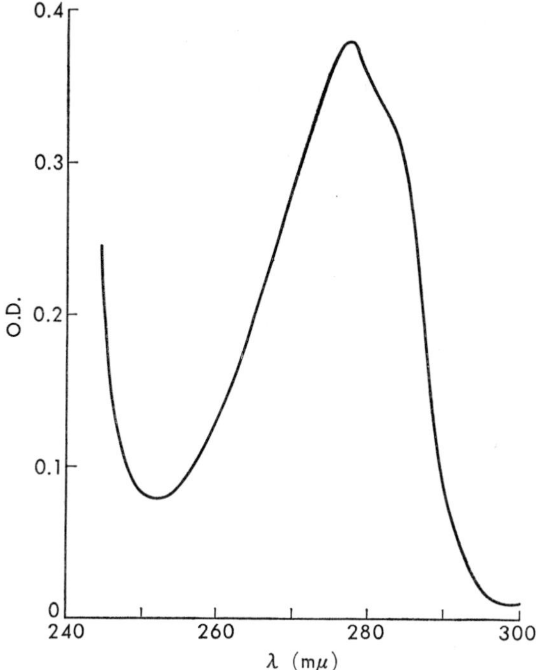

Fig. 1. Ultraviolet absorption spectrum of a tropomyosin solution. Extinction coefficient $\varepsilon^{1\%}_{277m\mu}=2.5$ for a path length of 1 cm. C_p : 1.5 mg/ml, pH 7.0, 5 mM phosphate buffer.

sulfate. Since the isoelectric point of tropomyosin is around pH 4.5, isoelectric precipitation is used to separate proteins that have different isoelectric points. It must be noted that the precipitation should be done in a concentrated salt solution to remove nucleoproteins and troponin strongly bound to tropomyosin. The second step, salting out by ammonium sulfate, is to take the precipitate between approximately 47 and 55% saturation. Actually, this step seems to be essential, because other proteins such as myosin and actin precipitate out at lower than 40% saturation. It is also important to carry out the salting out in the presence of salts at a neutral pH. Usually, three repetitions of the above two processes are sufficient to obtain the purified tropomyosin, although the criteria for purification of tropomyosin may be ambiguous since the protein has no specific biological function by itself. The tropomyosin polymer cannot be spun down even under a high gravitational field, so the polymerization property used for the purification of actin is not available for tropomyosin.   In addition, troponin or nucleoprotein may bind to the tropomyosin polymer, even if the separation of polymers by centrifugation could be done.

One of the criteria for examining the degree of purification is the ultraviolet spectra of solutions. Since tropomyosin contains no tryptophan residue, which has a strong absorption at 290 mμ (see the following section), the spectrum should be that for tyrosine with a maximum at 277 mμ (Fig. 1). If some of the nucleoproteins or troponin are contaminated in the preparation, absorption at around 250–260 mμ or at 290 mμ appears. A preparation that has gone through the purification process only once still has a strong absorption near 250 mμ, suggesting that nucleoprotein has been contaminated. Since the molar absorption coefficients for both nucleotide and tryptophan are quite large, this method would detect contaminants effectively.

The crystallization of a protein is often thought to indicate the isolation of purified proteins. Crystals of tropomyosin, however, which are grown in the presence of salts, about 0.2 M at pH 5.6, do not seem to represent the isolation of the purified protein, since native tropomyosin could be crystallized under the same condition (*10*) and also crystals of the same shape could be grown in the presence of actin (*11*). The preparation reported by Bailey is likely to contain native tropomyosin, judging from micrographs and their electron micrographs (*40*).

The other methods, such as column chromatography and gel electrophoresis, are applicable to the protein, but there seems to be no established method for purification of tropomyosin other than Bailey's procedure.

PHYSICOCHEMICAL PROPERTIES

1. Polymerization-depolymerization
When a solution of the precipitates obtained by salting out with ammonium sulfate is dialyzed against neutral water, the solution becomes very viscous. When the salts are added to the solution, the viscosity falls down quickly. This indicates polymerization-depolymerization of the tropomyosin molecules. The dependence of the viscosity on a salt concentration is shown in Fig. 2. The viscous feature of a tropomyosin solution differs in the absence of rigidity from a solution of actin or

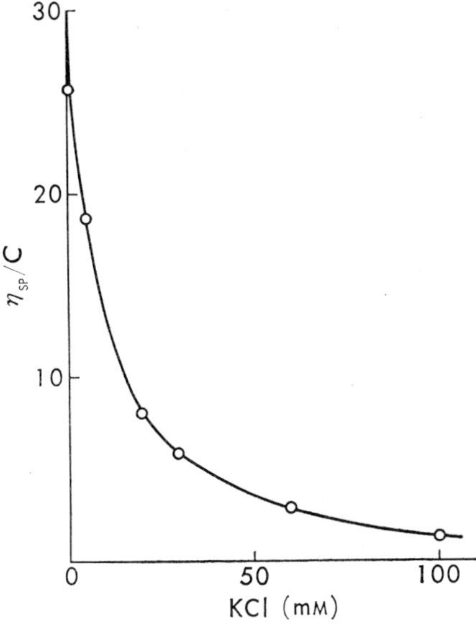

Fig. 2.   The dependence of the viscosity of a tropomyosin solution on a salt concentration.   Protein concentration :   1.5 mg/ml, pH 7.0.

native tropomyosin which shows flow birefringence even at a low shear rate.   The difference may be due to the absence of a comparatively strong network structure in tropomyosin solutions.

The polymerization of tropomyosin depends upon other environmental conditions: pH, temperature (*12*), and protein concentration (*13*), in addition to a salt concentration.   The dependence of viscosity on pH is shown in Fig. 3.   At a higher pH than 9.8, the tropomyosin polymer depolymerizes sharply, giving rise to a nonviscous solution, irrespective of temperature and salt concentration.   Because this phe nomenon is reversible, some ionizable groups may be responsible for polymerization.

On the other hand, it is not easy to derive direct information on the molecular structure from the results of the temperature dependence shown in Fig. 4, since the structure necessary for polymerization would be altered by temperature, the change being indicated in the measurement of ORD.   At any rate, reversible depolymerization occurs in the temperature range from 30 to 50°C, and the transition temperature depends on the pH of the solution (*12*).   It is noted that two peaks are

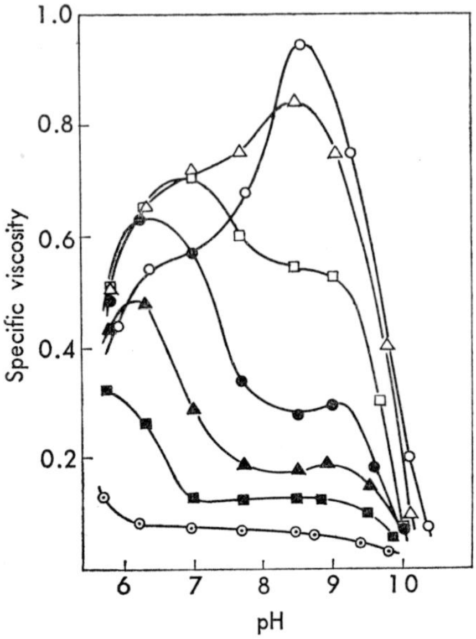

Fig. 3. Dependence of specific viscosity on pH as a function of temperature. Protein concentration: 1 mg/ml, KCl 10 mM, buffer 10 mM; ○, 10°C; △, 20°C; □, 25°C; ●, 30°C; ▲, 35°C; ■, 40°C; ⊙, 45°C. Phosphate buffer was used below pH 8.5 and borate buffer above pH 8.5.

observed in Fig. 3, one near pH 6.5 and the other at pH 8.5, suggesting two different kinds of effects of pH on polymerization. Recent results on the pH dependence of fluorescence at various temperatures indicate some structural changes with increasing temperature near the lower pH (*14*). The degree of polymerization as a function of pH and temperature has been investigated not only by viscosity but by other methods (*13*, *15*).

In order to show the dependence of polymerization on a protein concentration, several experiments have been carried out, *i.e.*, viscosity, electric birefringence (*15*), flow birefringence (*16*) and light scattering (*13*). All the results obtained by the use of the various methods indicate that depolymerization occurs with a decreasing concentration of protein. Among the quantities obtained from these experiments, those from light scattering are suitable for thermodynamical analyses. Apparent molecular weights measured by light scattering are illustrated as a function of the protein concentration in Fig. 5. With dilution, the molecular

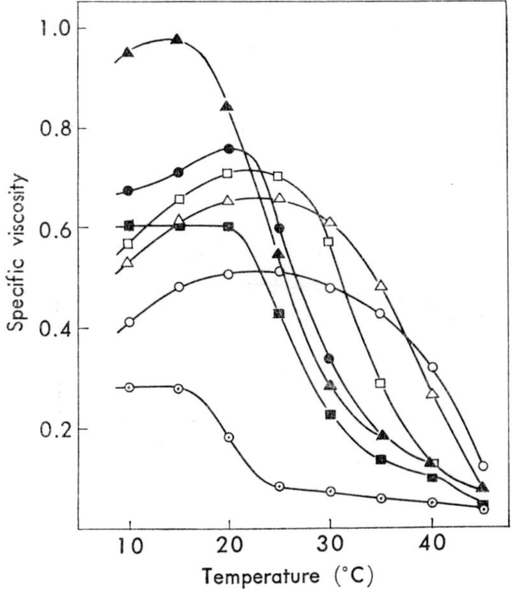

Fig. 4. Temperature dependence of specific viscosity as a function of pH.
○, pH 5.80; △, 6.30; □, 7.00; ●, 7.72; ▲, 8.50; ■, 9.50; ⊙, 10.0.
Replots of the data in Fig. 3.

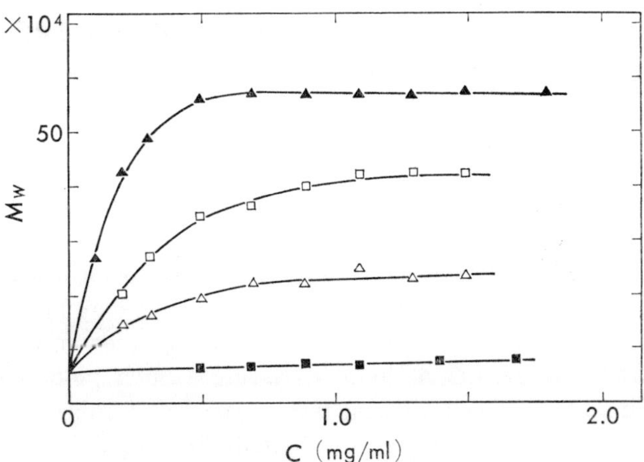

Fig. 5. Concentration dependence of apparent molecular weight on a salt concentration. Phosphate buffer: 5 mM, pH 7.0. ▲, 30 mM KCl; □, 60 mM KCl; △, 100 mM KCl; ■, 1000 mM KCl.

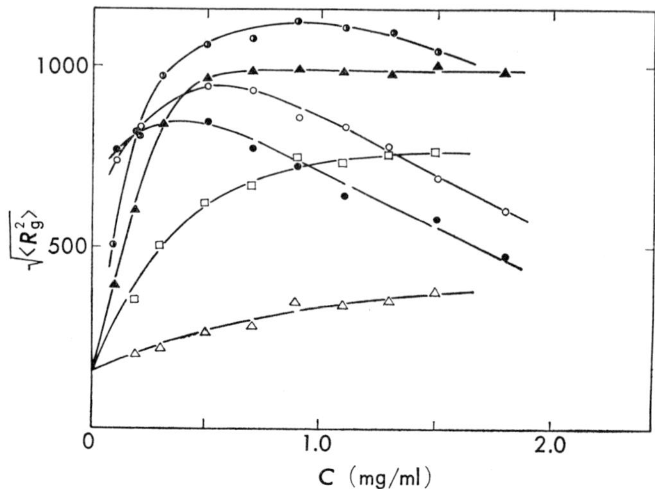

Fig. 6. Concentration dependence of R_g as a function of a salt concentration. Phosphate buffer: 5 mM, pH 7.0. ●, 0; ○, 5 mM KCl; ◑, 20 mM KCl; ▲, 30 mM KCl; □, 60 mM KCl; △, 100 mM KCl.

weight, which is dependent on pH and the salt concentration, decreases to that of a monomer. The corresponding relation of the radius of gyration *versus* the protein concentration is shown in Fig. 6. The existence of tropomyosin polymer is illustrated, for instance, by the radius of gyration obtained from an angular dependence of scattered light (Fig. 6), or more directly by electron micrographs of the polymers.

However, the determination of an averaged molecular weight at a finite protein concentration is not easy without some assumptions for the system because the measured quantities may be influenced by inter-molecular interactions. According to an early hypothesis given by Bailey, rod-shaped molecules aggregate end-to-end to form long fibers which yield a viscous solution (*17*). The hypothesis may be examined by comparing the molecular weight with the corresponding length estimated from the results of light scattering. Such plots are given in Fig. 7, where both axes represent the averaged molecular weight and length of molecules obtained experimentally by the light scattering. Since the solution can be regarded as a mixture solution of polymers that have various degrees of polymerization, the molecular weight-length relationship should be analyzed according to the averaging procedures

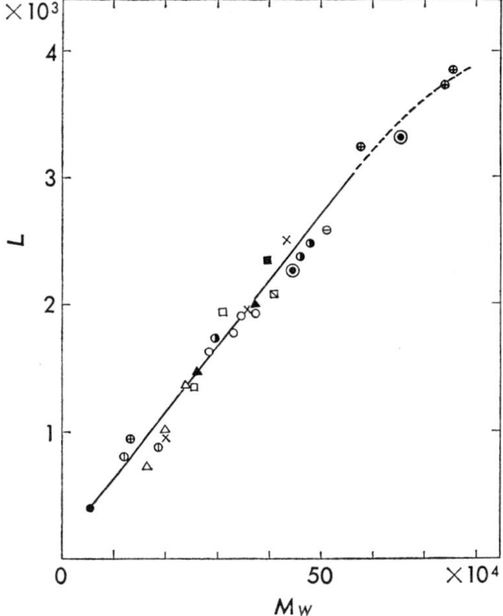

Fig. 7.   The relation between the molecular weight and the length. ⊙, KCl≑ 30 mM, pH 7.0 ; ×, KCl ≑ 60 mM, pH 7.0 ; △, KCl ≑ 100 mM, pH 7.0. Under 30 mM KCl, 30 mM buffer ; ○, pH 6.5 ; □, pH 7.0 ; ▲, pH 7.5 ; ◐, pH 8.1 ; ⊖, pH 8.8 ; ◻, pH 9.3 ; ■, pH 9.8. ●, monomer.

dependent upon the measured quantity. The molecular weight obtained by the light scattering technique is a weight average and the radius of gyration, a $z+1$ average (two orders higher than the weight average), so that a linear relationship between the molecular weight and the length of rod derived from the radius of gyration shown in Fig. 7 does not always represent the end-to-end aggregation of the molecule. The increment experimentally obtained is 290 Å per molecule (M=55,000), when the molecule is assumed to be rod-shaped. Therefore, the postulate of an end-to-end polymerization without side-by-side interaction seems to be improbable. A more detailed analysis based on polydispersity of molecular weight in solutions is necessary as described below.

2.   *Reversible equilibrium of tropomyosin polymers*
Since tropomyosin exhibits reversible polymerization phenomenon, monomeric units must combine with each other by some noncovalent

bonds under suitable circumstances. Therefore, the phenomenon may be treated as a problem of establishing equilibrium between monomer and polymer in a system where solute molecules are able to react with each other with the equilibrium constant K. It is assumed that the attachment of monomer to monomer gives a dimer, and generally, the addition of one monomer to i-mer produces $i+1$-mer, i.e., $M_1+M_1 \rightleftarrows M_2, \ldots, M_1+M_i \rightleftarrows M_{i+1}, \ldots$, where M_1 denotes a monomer and M_i generally denotes an i-mer. When monomer molecules of the total number of N exist in a volume, V, and we apply the mass action law to each chemical equilibrium, $M_i+M_1 \rightleftarrows M_{i+1}$, a number concentration for i-mer, the number of i-mer, N_i, divided by V or $C_i = N_i/V$, may be written as

$$C_i = \prod_{j=1}^{i-1} K_j C_1{}^i, \tag{1}$$

and therefore,

$$\sum_{i=1} iC_i = \sum i \prod_{j=1}^{i-1} K_j C_1{}^i = N/V. \tag{2}$$

On assuming that K_j has the same value, K, regardless of j, the distribution of i-mer in a given concentration, and average values of various degrees can easily be calculated by this distribution: e.g., $C_i = (KC_1)^i/K$.

As described before, a molecular weight obtained by the light scattering technique is a weight average, and the length from angular dependence is a $z+1$ average. Using the distribution, concentration dependences of the molecular weight and length are expressed in a power series when $KC \ll 1$ ($C = NM/V$, where M is the molecular weight of the monomer),

$$M_W/M = 1 + 2KC - 2(KC)^2 + \cdots, \tag{3}$$

$$\langle L^2 \rangle^{1/2} = L_1 \{1 + 2a(a+2)KC - \cdots\}, \tag{4}$$

where M_W is the weight average molecular weight, L is the length of the polymer, L_1 is the length of the monomer, and a represents the overlapping ratio in a form $L_i = L_1(1 + (i-1)a)$. A value of 0.35 for a is obtained from the slope of the line determined by the relation between the molecular weight and the length shown in Fig. 7. This implies that approximately three molecules are required for an increment of the monomer length, about 400 Å, in the tropomyosin polymer, that is, the side-by-side aggregation in addition to the end-to-end interactions is involved in polymerization. In fact, no thin long fibers could be found under the electron microscope. It is noted that the present analysis does not necessarily mean that the molecules should overlap by two-thirds of the length in growth to fibrous form; *e.g.*, it is possible that the polymers made up of monomers connected end-to-end might be associated side-by-side, resulting in a long fiber. In order to decide which is the case, further studies are necessary.

There is another deduction that can be derived from the dependence of molecular weight on the protein concentration. The equilibrium constant, K, at a given salt concentration, can be estimated from the curves in Fig. 5 by the use of Eq. 3. Once K is determined, the free energy of association per mole of the molecule can be deduced by $-RT \ln K$. The free energies thus evaluated at 1 M and 0.1 M KCl are -4.5 kcal/mole and -5.6 kcal/mole, respectively. Therefore appreciable attraction seems to be present between tropomyosin molecules. In view of the decrease in electrostatic interaction at a high salt concentration and the temperature dependence of polymerization shown in Fig. 4, it seems that some hydrophobic residues play a role in polymerization.

3. *Tropomyosin as a polyelectrolyte*

The strong dependence of depolymerization on a salt concentration and pH suggests the important role of ionizable groups of tropomyosin in the intermolecular interaction. The results of hydrogen ion titration experiments show that the titration curves have no abnormality for different degrees of polymerization except for the isoelectric region and that the additivity rule for the activity of the counter ions may be applied to the tropomyosin solution, suggesting that tropomyosin behaves as a polyelectrolyte (*18*). Also, the analysis of titration results supports the presence of the side-by-side interaction (*18*). Therefore, the role of charged groups and their distribution on the molecule seem to be essential for interaction between tropomyosin molecules. Since electrostatic interactions are greatly influenced by the media, *e.g.*, salt ions, and the dielectric constant, the polymerization must be dependent on the electrolytic property of the solvent.

In accord with the above consideration, the lowering of the dielectric constant by the addition of an organic solvent promotes polymerization. In addition, a large permanent dipole of 400 debye for a monomer was obtained by measuring the Kerr effect (*15*). A simple illustration of the polyelectrolytic behavior may be the dependence of viscosity on the ionic strength of divalent cations. When divalent cations such as Mg^{2+} or Ca^{2+} are added to tropomyosin solutions, depolymerization occurs at a lower concentration of the divalent cation because of the effect of the double charge. However, the further addition of ions gives rise to precipitation, showing some specific interactions with the protein. A strange property that makes it impossible for tropomyosin polymers to be centrifuged down may be accounted for in part by the charge effects of the polyelectrolyte in addition to the concentration dependence of the association-dissociation equilibrium.

4. *Monomeric unit of tropomyosin*

Since tropomyosin is polymerizable depending on environment and has the capacity to interact with other proteins, special care is required in the determination of monomeric unit as well as purification of the protein. The decontamination of troponin has been recognized as an important procedure to be performed after native tropomyosin has been found. Reagents have been used to avoid oxidation of thiol groups (*19, 20*) that might give rise to the aggregation of the molecules, thus resulting in a disturbance of the determination of molecular weight. The molecular weights obtained so far are from around 55,000 (*13, 17, 21*) to a higher value of about 70,000 (*22*), which has been reported recently. One of the reasons for the discrepancy may arise from the preparation of the materials; *e.g.*, some may contaminate troponin. Another reason may be due to the aggregative property which complicates the determination of the molecular weight of the monomer. A value of around 60,000 for monomeric units in aqueous solutions might be a probable one, but it is important that further experiments on the purified protein taking the subunit structure into account (see later) are carried out.

The high content of the helix in the tropomyosin molecule, detected by optical rotatory dispersion measurements (*23*), suggests a rigid rod-like molecule. Electron microscopic observation also shows the elongated shape having a length of about 400 Å. Assuming that a molecule of the molecular weight 6×10^4 is composed of a single α-helix which

Fig. 8a.   Electron micrograph of tropomyosin monomer.   KCl: 3 M.   Magnification ×120,000.

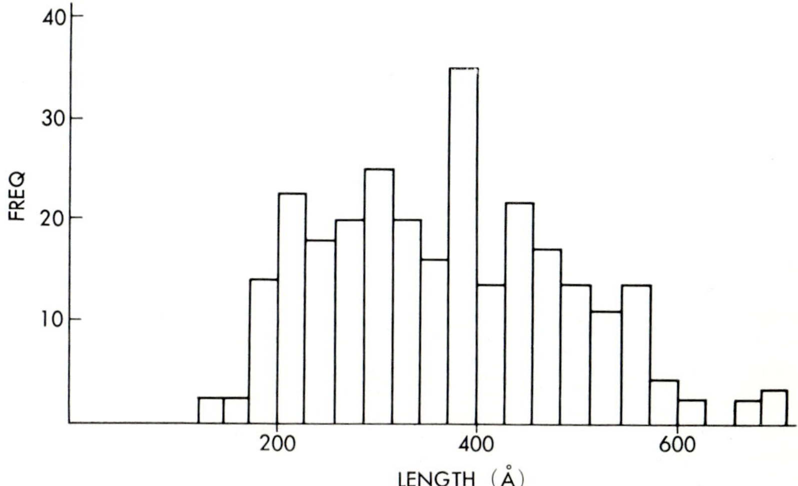

Fig. 8b.   The distribution of the length of monomers found in the photograph (Fig. 8a).

has the periodicity of 5.1 Å per 3.6 residue observed in an X-ray diffraction pattern, the length of the molecule would be about 800 Å. Accordingly, the molecule seems to consist of two α-helices whose lengths are approximately 400 Å each, and the coiled-coil model proposed by Crick (*24*) is quite a reasonable one.   Preliminary observations of monomeric units in the electron microscope show that the molecules have an average length of 400 Å over the range from 200 to 600 Å (Fig. 8b) with a width of about 20 Å and that the shape is not always straight but frequently bent or curved, suggesting the existence of a loose region in the molecule (Fig. 8a).   This flexibility may be correlated with an easy attack of proteolytic enzymes on the molecule described later. Electron microscopic studies on the tropomyosin monomer with the mica-replica technique have shown similar pictures (*25*).

There are several investigations which show the split of the molecule into two subunits in the presence of denaturing reagents with mercaptoethanol, a high concentration of urea (*26, 27*), guanidine HCl (*28*) and sodium dodecyl sulfate (SDS) (*29*).   Unfortunately, the separation into subunits in aqueous solutions has not yet been successful. Although the subunit structure of tropomyosin seems to be very probable, we cannot draw a definite conclusion, for the moment, especially because the determination of molecular weight in the denaturing media involves several problems in obtaining a real molecular weight, *e.g.*, an estimation of a partial specific volume.   The simple technique of gel electrophoresis in the presence of SDS has been applied to tropomyosin. The result gives a value of about 37,000 (*29*).   This technique also should be examined further.

TABLE I

Amino Acid Composition of Tropomyosin

Lysine	107	Glycine	12.5
Histidine	5.5	Alanine	110
Ammonia	64	Half cystine	6.5
Arginine	42	Valine	27
Aspartic acid	89	Methionine	16
Threonine	26	Isoleucine	30
Serine	40	Leucine	95
Glutamic acid	212	Tyrosine	15
Proline	1.7	Phenylalanine	3.3

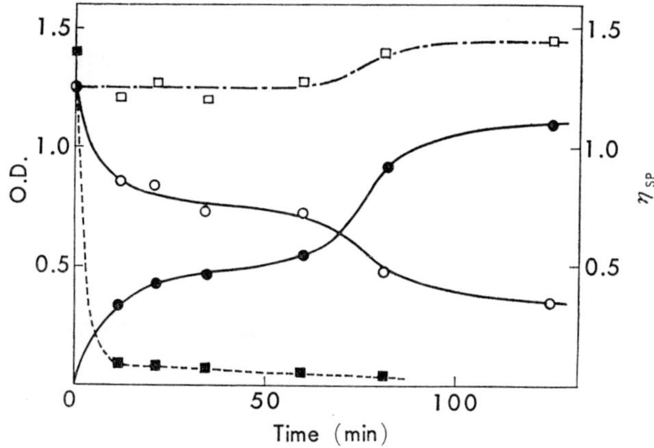

Fig. 9. The amount of fragments in the precipitate (○) and in the supernatant (●) expressed in optical density at 277 mμ at various time intervals after the addition of trypsin. The addition of both curves is shown by a chain line (□). The viscosity for solutions of the precipitate (■) is represented by a broken line.

BIOCHEMICAL STUDIES OF TROPOMYOSIN

The amino acid composition of tropomyosin is shown in Table I. The main feature of its composition is the presence of a large number of ionizable groups, particularly glutamic acid, though the analysis could not distinguish glutamic acid or glutamine. The high content of hydrophilic residue may be related to the helical structure of the protein. Since the tropomyosin of a high purity has been available, some corrections may be necessary, presumably on the content of lysine, cystine and proline, judging from the preliminary amino acid analyses done on the purified tropomyosin. It has been reported that C-terminals are isoleucine and/or serine (30, 31). Recent examinations using carboxypeptidase A show the existence of two possible terminals, one -Leu-Ileu, and the other -Thr-Ser (52), supporting the suggestion that tropomyosin really consists of two different peptide chains. As for N-terminals, it has been reported that glutamic acid (32) and the unknown residue, probably acetylated, most likely make up the N-terminal residues (33). These results are not so reliable, however, because the yield of the N-terminals was low and there is another experiment in which different terminals were found (35).

Despite the high helical content, tropomyosin is easily attacked by a proteolytic enzyme such as trypsin, so that the technique of limited proteolysis can be used for this protein in order to identify a structurally loose region in the molecule (*34*).   As shown in Fig. 9, peptide chains are cleaved by trypsin into two parts, one that can be precipitated out at an isoelectric point of pH 4.6, and the other which is soluble at this pH.   Interestingly, the precipitates can be dissolved at a neutral pH; and the solution is not viscous, indicating a loss in polymerizability by cleavage.

Analyses of the precipitate and the supernatant by sedimentation and gel filtration show that the precipitate contains relatively homogeneous large fragments, and the supernatant, various sizes of small fragments. The large fragment still has a structure (60% helical content) similar to tropomyosin, and a value of 37,000 for the molecular weight is obtained by sedimentation (both Archibald and sedimentation equilibrium) experiments.   The results suggest that the cleavage of the peptide bonds occurs in the region located about one-half to two-thirds of the way from the end, while the smaller part of the molecule is easily digested further by trypsin, resulting in acid soluble peptides.   Therefore, the structure of tropomyosin is presumably constituted of two parts connected at some loose region(s) which is accessible for enzymic attack.

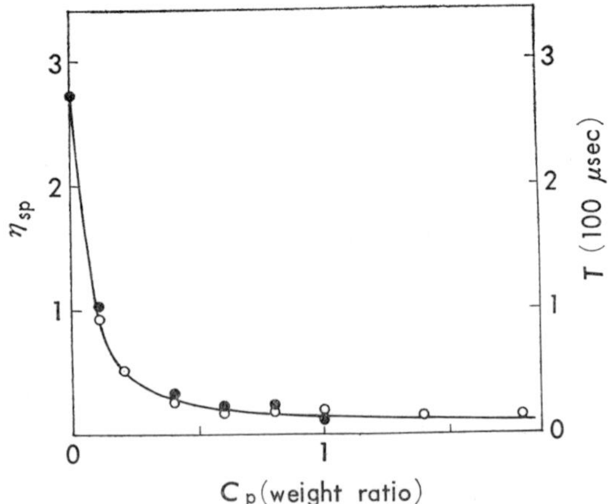

Fig. 10.   Effect of the large fragment on polymerization of tropomyosin.   (O) Specific viscosity.   (●) Relaxation time from Kerr effect.   Tropomyosin concentration :   1.5 mg/ml, 0.007 M phosphate buffer (pH 6.9) at 25°C.

It may be of interest to mention that the fragment is able to de-polymerize tropomyosin polymers as shown in Fig. 10. With the addition of the fragment, the viscosity of the tropomyosin solution falls rapidly. The result indicates that the fragment has a structure that interacts with the tropomyosin monomer but has little interaction with the fragments.

Several studies on the thiol groups in tropomyosin have been done (7, *19, 35, 36*) because a disulfide bridge would make it possible to connect the chains on the coiled-coil model. At the present stage, it is not clear yet whether the disulfide bridge is contained in the molecule or not.

CRYSTALS OF TROPOMYOSIN

Tropomyosin is probably the only fibrous protein that can grow into three dimensional crystals. In order to elucidate the molecular structure, this crystal would be available for X-ray analysis, the technique that has been employed successfully to determine the tertiary structures of several globular proteins (*37*). Another technique to identify the molecule, electron microscopy, can also be applied for the present purpose (*40*). Actually, we can observe the fibrous structures of F-actin, myosin A, tropomyosin and the monomers of these molecules under an electron microscope (*25, 38, 39*).

However, the resolution of the electron microscope, and, especially, the noise that arises from the grains of staining materials on the background are serious problems in obtaining a distinct image of the molecule. With the hope of reducing the noises from the background, electron microscopic studies on tropomyosin crystals have been begun because the periodic arrangement of the molecules in a crystal may be advantageous for distinguishing the molecular structure in detail (*11*).

1. *Crystallization*

A solvent condition for crystallization does not seem to be critical in the salt range from roughly 0.1 to 0.4 M and the pH range from 5.9 to 5.1 (*11*). When a higher salt concentration or a higher pH is chosen for crystallization, the growth of the crystal is slow, *e.g.*, several weeks. The best condition may be 0.2 M KCl at pH 5.6. The presence of the divalent cation Ca^{2+} or Mg^{2+} affects the crystal form, *e.g.*, paracrystals

                                          T. OOI and S. FUJIME

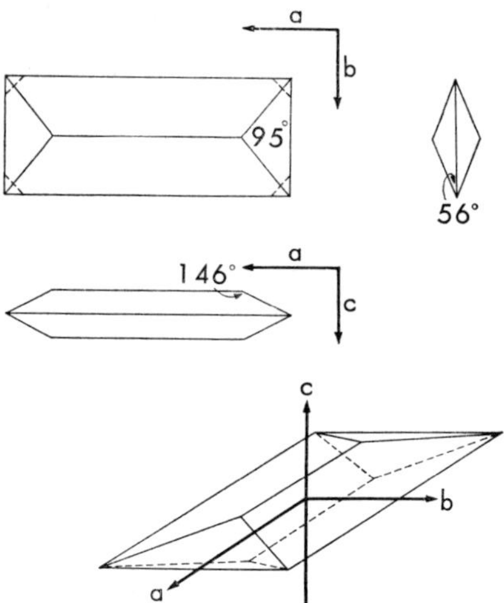

Fig. 11.   Shape of tropomyosin crystal.

are formed at the neutral pH (*41*).   In the temperature range from 0°C
to room temperature, significant effects on the crystal formation have
not been observed, so usually crystallization is performed in a cold room
to prevent germination.

As described before, crystallization does not imply the purity of
the protein, since other muscle proteins can be incorporated into crystals.
Native tropomyosin, or tropomyosin with troponin, crystallizes in a
different shape (*10*), whereas crystals of the same shape can be obtained
with the addition of actin.   The use of purified tropomyosin gives
rise to a crystal of the shape shown in Fig. 11.   Angles measured under
the microscope are demonstrated in that figure also.   The crystal is
a fragile thin flake; and, occasionally, large crystals with cracks running
through the center peak of the crystal have been found.   Sometimes
a crystal is grown with daughter crystals in it.   The crystal shape shown
in Fig. 11 is a typical one for tropomyosin.   This is not a unique shape:
we found different types of crystals under the electron microscope
depending on the crystallization conditions.   However, any single
crystal having a well-defined crystal habit other than the one shown in
Fig. 11 has not been observed.   The mixed crystal of tropomyosin and
actin has a very rigid structure in comparison with the crystal formed
from only tropomyosin.

Fig. 12. An electron micrograph of the typical tropomycsin crystal. 200 Å network.   Magnification × 200,000.

2. *Electron microscopic observations*

A typical electron micrograph of the crystal after negative staining with uranyl acetate is shown in Fig. 12. Judging from the overall shape under low magnification, this structure corresponds to the crystal shown in Fig. 11. Since an electron micrograph is a two-dimensional image of the crystal, we have information only on the structure as projected onto a two-dimensional plane. The photograph shows that the crystal has a network structure composed of tetragons, each of which has longer sides of 220 Å in length and shorter sides of 180 Å. The tetragon, in which the angle between the longer sides is 82° and the angle between the shorter sides, 107° on the average, is neither a parallelogram nor a rhombohedron, so that the neighboring four units constitute a repeating unit of translation in the network structure. As a result, parallel zigzag lines in two directions constitute the network, crossing each other with an approximate angle of 75° or 105°. The longer edge of

Fig. 13.   An electron micrograph of the crystal by shadowing.   Magnification $\times 80,000$.

the crystal corresponds to the edge of the network that is perpendicular to the longer diagonals of the tetragons.   The thickness of the crystal is easily distinguished by a gradual darkening toward the center of the crystal due to the penetration of the staining solution.   It is not clear from this photograph whether the network consists of a three-dimensional cage or a piling up of two-dimensional networks.   A photograph of the crystal taken by shadowing is shown in Fig. 13, where the tetragons are recognizable.

$\rightarrow$Fig. 14.   Electron micrographs of the 200 Å network showing the edge of the crystal.   The crystal has been homogenized.   a : Magnification $\times 200,000$. b : Magnification $\times 240,000$.

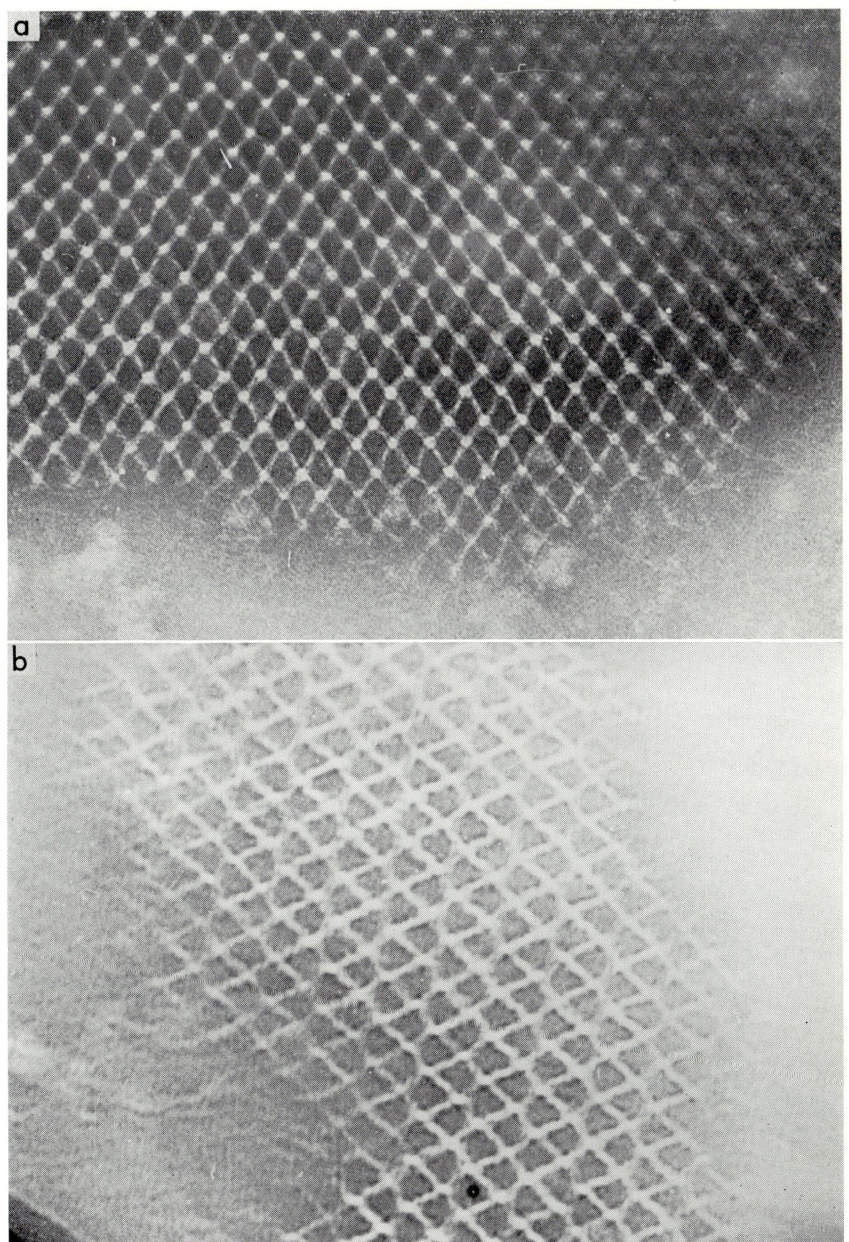

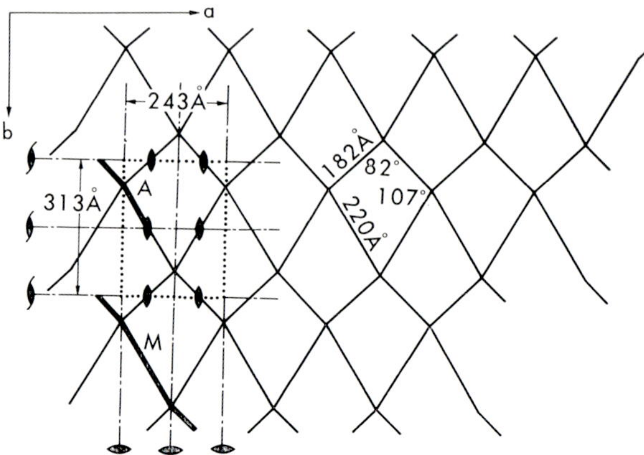

Fig. 15.   Crystal structure of the typical tropomyosin crystal.   Dimensions given in the figure are mean values.   A and M represent the asymmetric unit and the molecular unit, respectively.

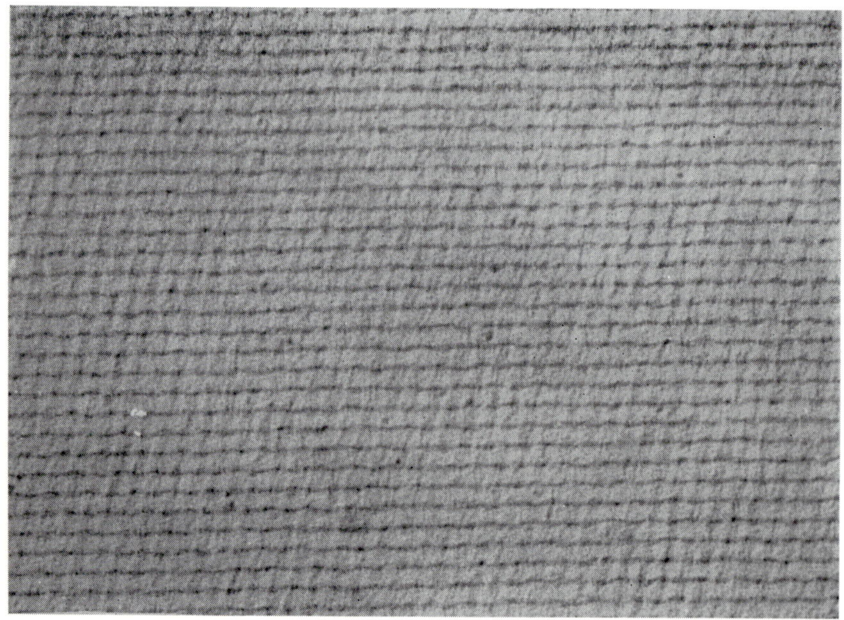

Fig. 16.   An electron micrograph of an ultrathin section of the crystal parallel to the *a-b* plane.   Magnification ×140,000.

When crystals are homogenized to break them down into pieces, small networks are found under the electron microscope as shown in Fig. 14. The photograph gives us further information on the structure. First, the edge of the network which corresponds to the longer edge of the crystal always ends at vertices of the longer sides of the tetragons, *i.e.*, both longer edges have the same network structure and therefore we cannot fill the network only by the translation of a repeating unit of four neighboring tetragons. Second, breaks occur along the other edges of the network leaving long and short tails alternately, suggesting the way the molecules are connected (see Fig. 15). Third, it is presumed that the crystal is formed by piling up the two-dimensional networks because the network structure would be broken if the crystal is a cage of the three-dimensional network.

According to the results obtained from the photographs, the structure of the network may be deduced to be as shown in Fig. 15. There are twofold screw axes of symmetry parallel to the *a*-axis, and two-

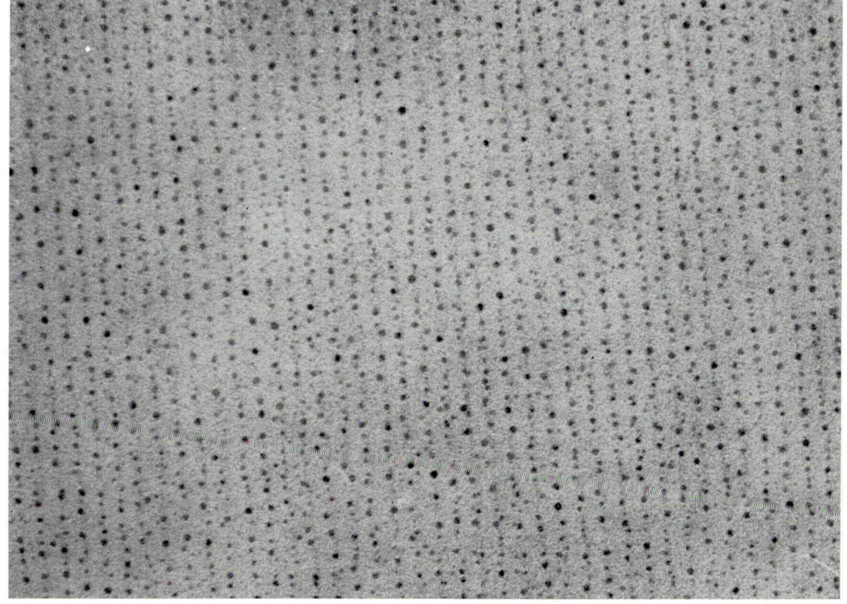

Fig. 17. A high magnification of the section perpendicular to the *a-b* plane. Dark spots of about 70 Å separation constitute the dark lines. Magnification ×140,000.

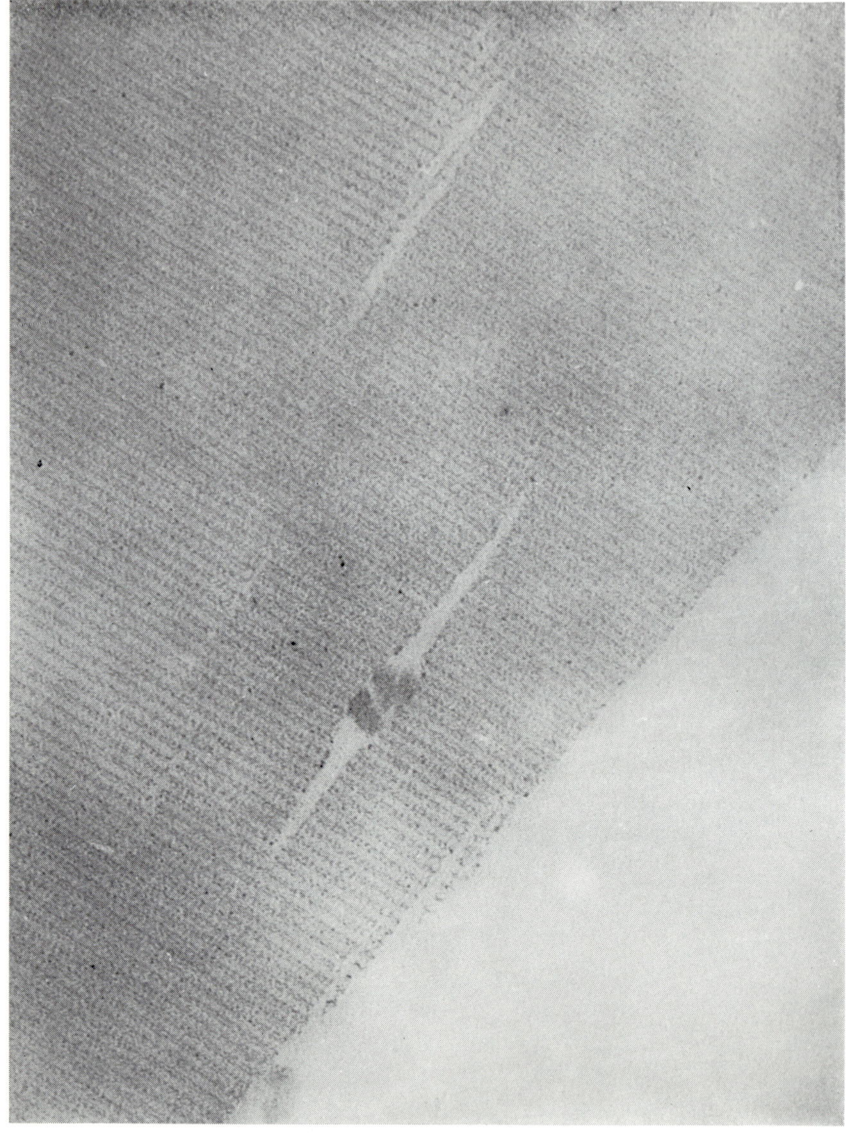

Fig. 18. An electron micrograph of an ultrathin section of the crystal perpendicular to the *a-b* plane. The distance between dark lines is 200 Å. The micrograph shows cracks running through perpendicular to dark lines. Magnification ×100,000.

fold axes along the *b*- and the *c*-axis. Therefore, the crystal belongs morphologically to a space group of $P2_122$, which includes the space group of $P2_12_12$ derived from X-ray diffraction studies (*42*), *i.e.*, symmetry about the *c*-axis cannot be identified whether twofold axis or twofold screw axis, since the electron microscopic observation is only concerned with the projection of the crystal onto the *a-b* plane. The unit cell, or the translational unit of symmetry, is shown in Fig. 15, with *a* dimensions of 243 Å × 313 Å. The asymmetric unit (*A*), one-half of the molecular unit (*M*), is taken from a midpoint of a shorter side to the midpoint of the adjacent longer side with a length of 200 Å. There are too possible ways in taking the asymmetric unit, but the arrangement of the asymmetric unit shown in the figure (the thick line in the figure) may be probable; *i.e.*, the units lie on the zigzag lines (*11*). The thickness of the line is about 20 Å and the separation into two or more thinner lines of approximately 10 Å thick has been observed. This

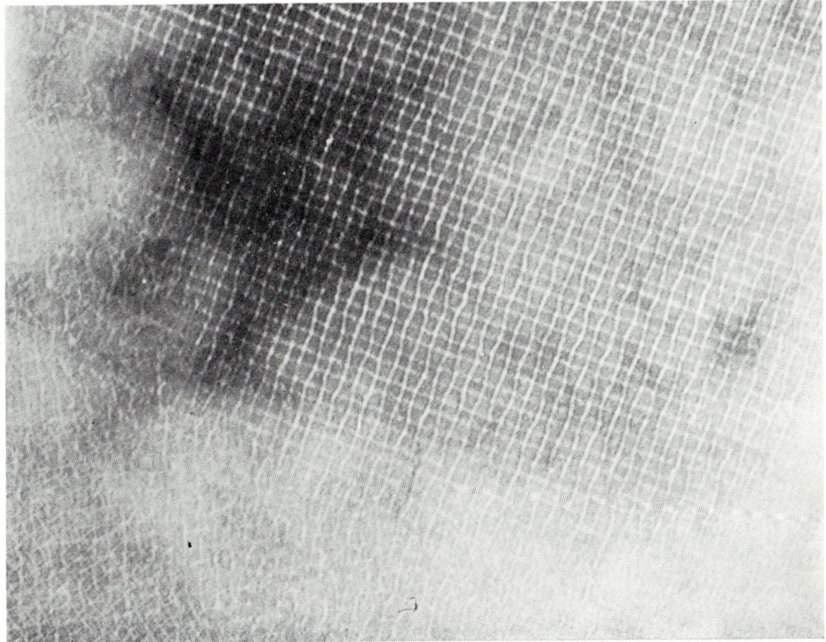

Fig. 19. Another form of 200 Å network. The network lattice is approximately square. See text. Magnification ×100,000.

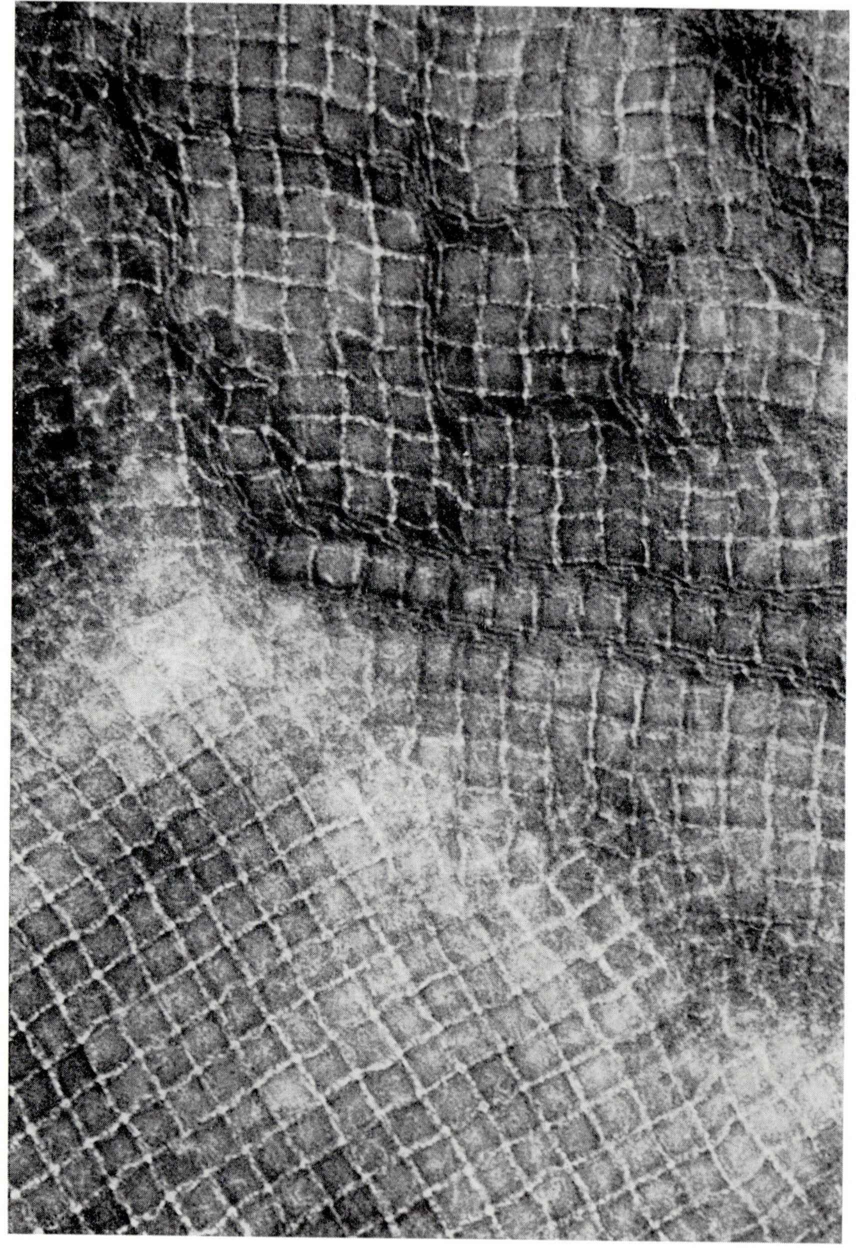

indicates that the line does not consist of a single strand of the tropomyosin polymer made by the end-to-end connection of monomers.

In order to see the structure along the *c*-axis, the sectioning technique is used after the fixation of the crystal with gultaraldehyde followed by molding in epoxy resin.   An electron mocroscopic picture of a section parallel to the *a-b* plane shows the same pattern as that in Fig. 14 (Fig. 16), and the photographs in Figs. 17 and 18 show the cross-section perpendicular to the *a-b* plane.   We can recognize dark lines that consist of fine dots with an approximate period of 70 Å running parallel to the *c*-axis with a separation of 200 Å.   There are dark dots at the ends of the lines, corresponding to the crystal edge.   Therefore, the crystal seems to be composed of a pile of planar networks rather than a three-dimensional cage.   This deduction is supported by the shape of the cracks in the crystal shown in Fig. 18, where we recognize the cracks running perpendicular to the dark lines.   The dark lines on both sides of the crack can be connected, one to the other, if the cavity is closed, suggesting a structural weakness in the crystal along the line perpendicular to the dark line.   Any crack parallel or skewed to the line has never been observed.

From the results described above, the crystal structure of tropomyosin is deduced as follows: tetragons with sides of 220 Å and 180 Å form a network structure and the networks pile up with a distance of about 70 Å to constitute the crystal.   In the network, the molecules in the line may be arranged in an antiparallel fashion if the molecule itself is not symmetrical because the molecular unit must have the symmetry.

3.   *Polymorphism of the crystal*

Crystals can be grown under different conditions from KCl 0.2 M, pH 5.6, although single crystals with a well-defined crystal habit are not obtained   The kind of ions seems to affect the crystal structure.   For instance, when an ammonium acetate solution is used as a crystallization solution, a similar but different network structure has been observed under the electron microscope.   The network has the same periodicity of 220 Å and 180 Å, but filaments cross each other at approximate right angle, or more exactly, the network is constituted of squares, rectangles

---

←Fig. 20.   An electron micrograph of the 400 Å network grown in the presence of PCMB.   Magnification ×150,000.

and trapeziums whose sides are 180 Å and 220 Å long, as shown in Fig. 19.

Another crystal having two dimensional structure has been found when the crystal was grown by the addition of PCMB to a crystallizing solution. Although this crystal can also be grown in other media, the addition of PCMB seems to be the best way for the preparation. The shape of the crystal under the microscope has not been identified yet. The network shown in Fig. 20 consists of a square lattice of 400 Å separation. Judging from the ends of the filaments and the breaks of the network clearly observed in Fig. 21, the molecules crossing over the connecting point would be the molecular unit of this crystal. The length,

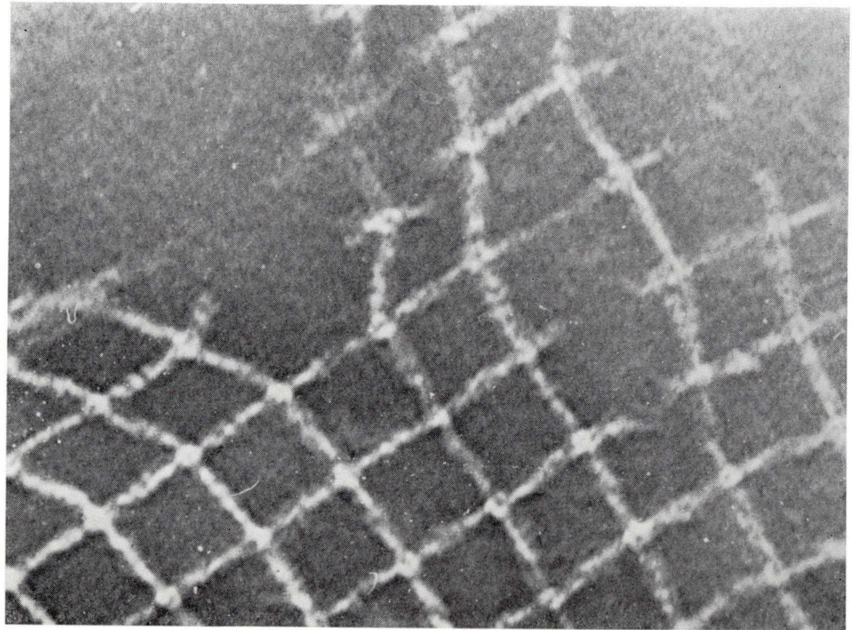

Fig. 21. An electron micrograph showing breaks in the 400 Å network. Connecting regions of molecules are seen. Magnification ×350,000.

→Fig. 22. (upper). An electron micrograph of the 400 Å network showing details in the arrangement of filaments. Magnification ×320,000.

→Fig. 23. (lower). An electron micrograph of the tropomyosin paracrystal. Band structure is seen. Magnification ×250,000.

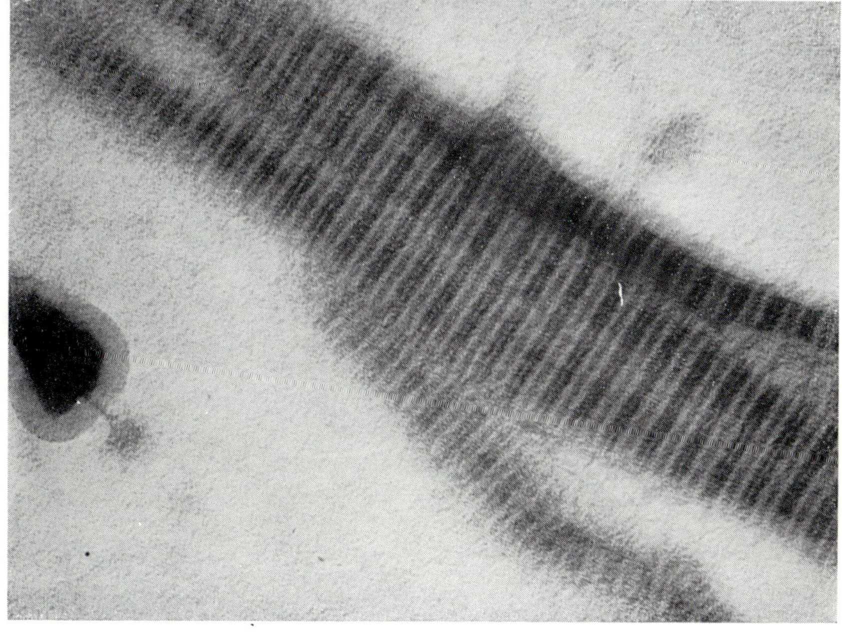

400 Å, corresponds to the length of a molecule, so that molecules are located from the midpoint of one side of the square to the midpoint of the adjacent side on the line.

Because of the two-dimensional structure and because of the 400 Å separation, the arrangement of the molecules can be observed in detail in the photograph shown in Fig. 22. The sides consist of bundles of two or more filaments, each of which is about 15 Å in width. The filament is probably a fiber from the monomer molecules. Another interesting feature of the filament is the four dots along the side of the square (Fig. 20), suggesting an 80 Å periodic structure. When the tropomyosin molecule is a coiled-coil with two α-helices, the periodicity expected is about 80 Å to 90 Å. Thus, the dots may indicate the coiled-

Fig. 24.   An electron micrograph of the hexagonal network.   Magnification × 100,000.

→Fig. 25.   An electron micrograph showing the polymorphic transition of crystals.   Magnification × 100,000.

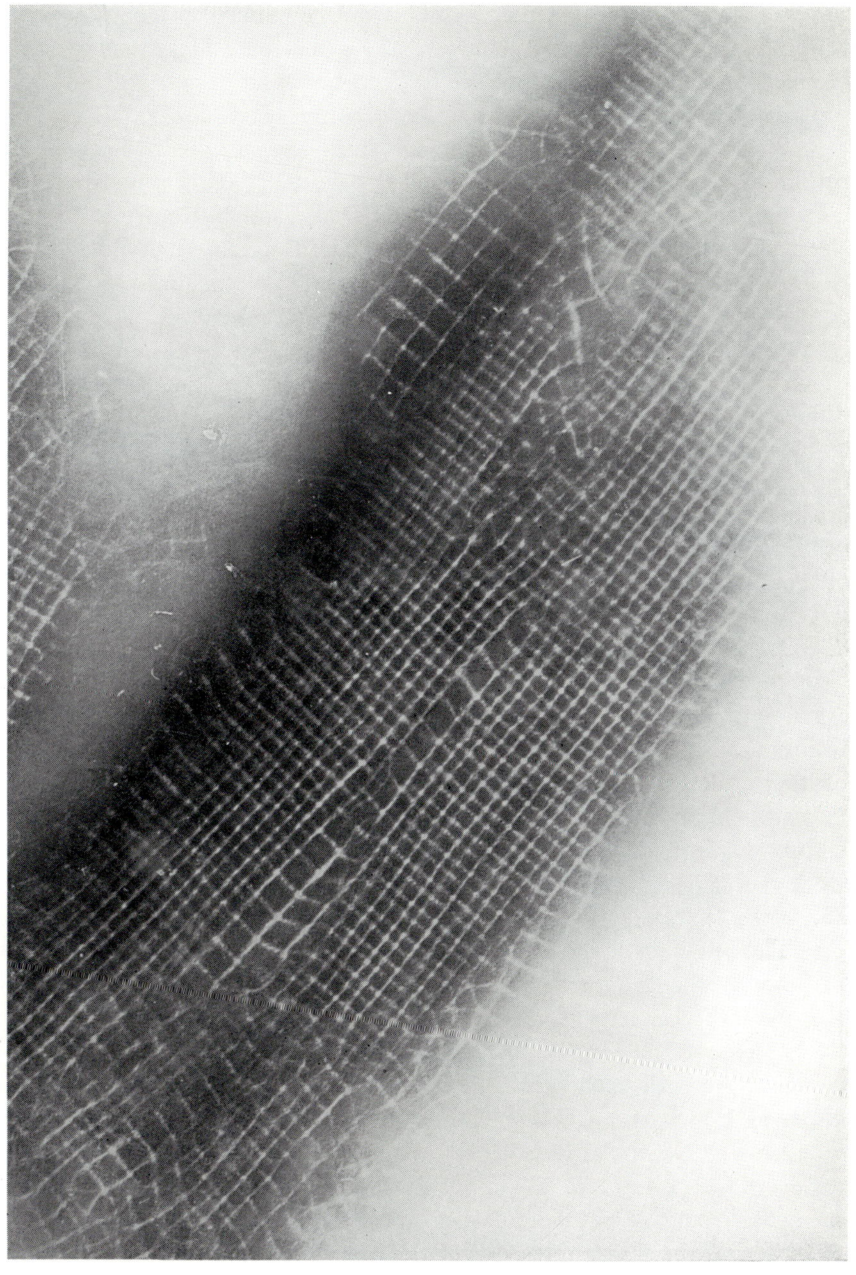

coil structure of the molecule. This is not conclusive yet because the winding of the filaments might give the same periodicity. However, the fact that the periodic structure is often observed on a single filament supports the coiled-coil model for the tropomyosin molecule (*43*).

The other form of crystal, paracrystal, is easily grown in several conditions, *e.g.*, in the presence of Mg^{2+} or Ca^{2+} (*41*) at a neutral pH. The general feature of this paracrystal is a band structure with the alternate repetition of white and dark bands, the periodicity of which is 400 Å, although the width of the bands is different from crystal to crystal depending on the crystallizing condition. An example is shown in Fig. 23. The paracrystal has a more dense structure than those described before and the molecules are packed compactly along the fiber axis. The result also shows that the length of the molecule is approximately 400 Å. The dark and bright regions may correspond to the density of the materials along the paracrystal fiber. Sometimes a crystal with a two-dimensional hexagonal lattice with a side of 400 Å has been found, as shown in Fig. 24. So, we have found several kinds of tropomyosin crystals.

4. Polymorphic transition between various crystal forms

The crystals of tropomyosin are polymorphic depending upon crystallization conditions, and, therefore, it is expected that the transition from one form to another may occur with a variation in the environment of the crystal. A transition has been observed under the electron microscope when the crystals were immersed in another solvent, *e.g.*, water. Transient states from a 400 Å network to a 200 Å square network or to a hexagonal network are seen in the photograph shown in Fig. 25.

The polymorphism of a tropomyosin crystal may be due to the variation in the structure of the molecule, *i.e.*, some sites of the molecule that have the capacity to interact with another molecule may vary with the environment. That is, at the condition that two parts of the molecule can interact with each other, a 200 Å network would be formed, and when only one part, in the center of the molecule can interact, then a 400 Å network would be formed. The difference in interacting sites arises conceivably from the modification of interacting site(s) by salt bridges,binding of divalent cation, the masking of a specific site and so on. In every case, both ends of the molecule cannot be distinguished from the other parts of the molecule except for the broken side, probably due to a linkage such as the connection of ropes without knots.   In some cases, the transition from a 400 Å square lattice to a paracrystal could be observed (*11*).

INTERACTION WITH OTHER MUSCLE PROTEINS

1. *Actin*

The increase in the flow birefringence of F-actin solutions with the addition of tropomyosin suggests an interaction between tropomyosin and actin.   Although the nature of the interaction has not been investigated in detail yet, filaments of both proteins seem to interact with each other (*44, 45*).   Electron microscopic studies were not successful in identifying the interaction because of the difficulty in distinguishing one filament from the other.   The other evidence is the formation of a crystal with actin in it.   The electron micrograph of the section of the rigid crystals shows that some aggregates seem to fill the cavity in the crystal (*11*). In addition, the electron micrographs of the crystal edge negatively stained do not show the network structure observed for the tropomyosin crystal, but some materials seem to fill the space of the network. Therefore, the rigidity may arise either from filling the spaces in the network structure with actin or from filling the cavities in the crystal with F-actin or both.   The study of this interaction may be of interest in the future.

2. *Troponin*

Troponin is strongly bound to tropomyosin, giving rise to native tropomyosin.   The binding of troponin can be detected by the increase in viscosity with the addition of troponin (*5*).   In contrast with the non-elastic property of tropomyosin solutions, the solution of native tropomyosin exhibits visco-elastic behavior.   The native tropomyosin can be crystallized as mentioned before.   Electron microscopic observation shows that the crystal of native tropomyosin consists of skewed lines of a 300 Å separation with some materials in the background as shown in Fig. 26.

In order to identify the location of the binding site of troponin with tropomyosin, electron micrographs of the crystal with a low troponin content were taken.   Figure 27 shows that the binding occurs at the midpoint of the longer side of a tetragon in the 200 Å network.   Therefore, the lines of a 300 Å separation correspond to the lines of the connecting midpoints of the longer sides of tetragons.   The formation of 300 Å lines is clearly seen in Fig. 27.   This binding was also observed on the paracrystal by using anti-troponin (46).

The binding of troponin onto a crystal of the 400 Å square lattice can be investigated by pouring a drop of the troponin solution onto crystals on a mesh.   The photograph shown in Fig. 28 illustrates the binding of troponin in a small amount.   Although surface tension might

Fig. 26.   An electron micrograph of native tropomyosin.   Magnification ×120,000.

→Fig. 27. (upper).   An electron micrograph showing the binding of troponin onto the typical tropomyosin crystal.   Magnification ×140,000.

→Fig. 28. (lower).   An electron micrograph of troponin binding onto the 400 Å network in a small amount.   Magnification ×120,000.

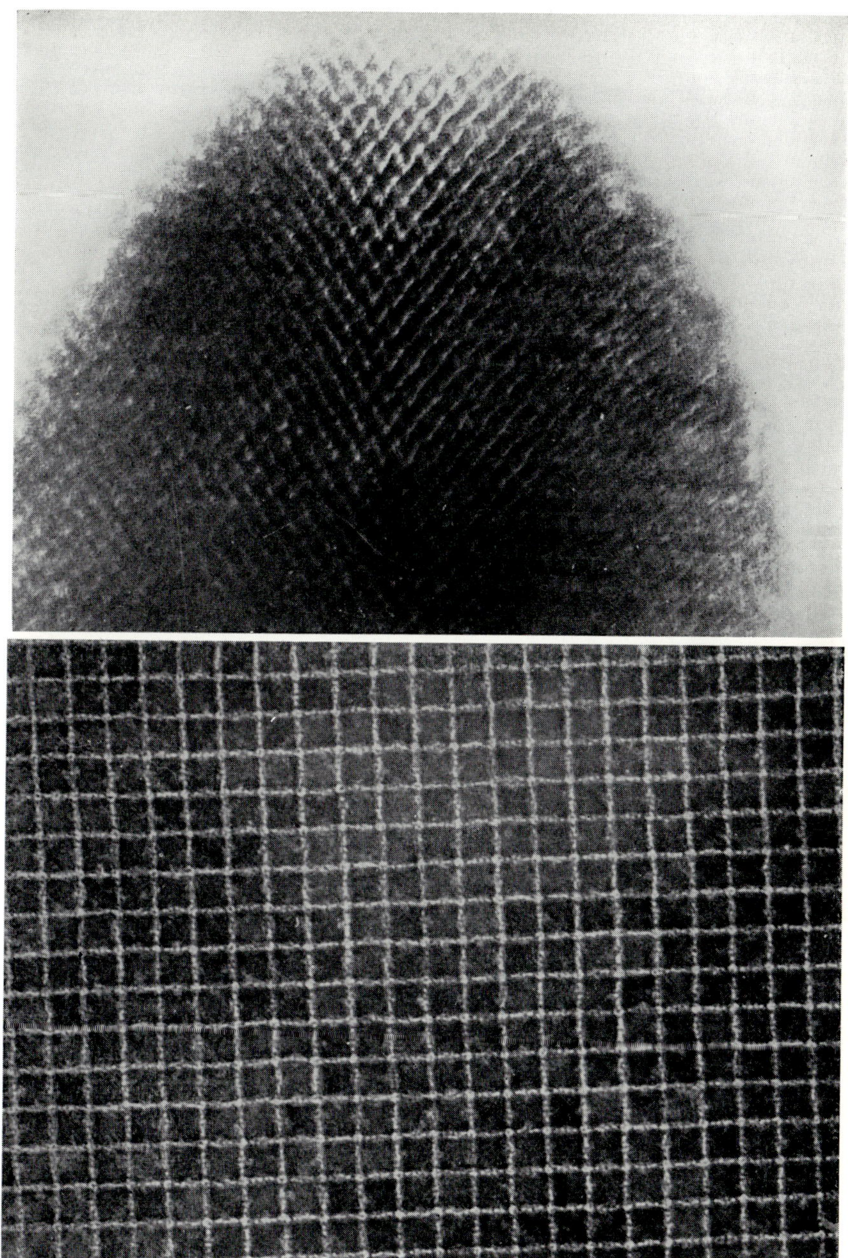

have an effect on the result, it appears that the binding site would be at the midpoint of a side or the connecting region of the molecules. This is different from the binding site on the 200 Å network. The reason for the discrepancy is not clear yet. The possible explanations are as follows: first, the binding site of troponin might be modified in the presence of PCMB; second, the apparent binding site might be an artifact due to surface effects; and last, the thin monomeric molecules might bridge over the side to cover the connecting region. The determination of the binding site seems to be important in connection with the studies on the structure and function of proteins in the thin filament.

3. *Other proteins*

The major component of the muscle proteins, myosin A, does not exhibit any binding to tropomyosin. However, some nucleoproteins must have a strong binding to tropomyosin because, during purification, a certain nucleoprotein binds so strongly that the repetition of isoelectric precipitation and ammonium sulfate salting out at a high salt concentration is necessary. There have been no studies along this line reported so far. Another component of muscle proteins, α-actinin, seems to interact with tropomyosin (*46*).

POSSIBLE STRUCTURE OF TROPOMYOSIN MOLECULE

The information obtained so far from physicochemical measurements and electron microscopic observations may allow us to build a tentative model of the molecule. The rod-like structure of the coiled-coil molecule is consistent with a high helical content and the shape of the molecule under the electron microscope. It has not been decided whether the two α-helices are parallel or antiparallel. At least, the molecule is asymmetric or it has a head and a tail, because a large permanent dipole of about 400 debye cannot arise from a symmetric molecule. If the molecule consists of a single peptide, the antiparallel arrangement is the only possibility.

For the two subunit model, a coiled-coil molecule composed of two parallel peptide chains is as equally probable as an antiparallel one.   From the view that a permanent dipole moment of an α-helix is proportional to the chain length due to the parallel arrangement of the dipole of the peptide, 3–4 debye, the antiparallel structure seems to be energetically stable because the great amount of repulsion between parallel dipoles would be too strong to overcome the attractive forces between the side chains that could have originated mainly from hydrophobic interaction.   The energy calculation of coiled-coil molecules shows more stability for an antiparallel chain (47, 48).   If we assume the antiparallel arrangement, the heterogeneous distribution of ionizable groups along the molecule would give rise to a large permanent dipole.   For example, two unit charges of the opposite sign with a separation of 100 Å would result in 480 debye; and the residual dipole from the main chains would become small enough by cancellation of the antiparallel dipoles.

Although the molecule has an elongated shape, some flexibility must be present, as shown by the network structure of the crystal and by the picture of the bent form of the monomer.   The cross points in the network and the arrangement of monomers in the crystal suggest that the loose regions would be located about 90 Å from both ends, where attractive sites between molecules seem to exist.   Also the middle part of the molecule may be some specific sites for interaction.   Of course, these deductions are based on the assumption that the sites are not influenced by pH, salt concentration, *etc.*   The assumption might be invalid if the charge distribution has a serious effects on the structure.

Biochemical studies of the limited proteolysis by trypsin show the presence of the loose region(s) in the molecule that leave a resistant core.   There would be some correlations between the specific sites observed in electron micrographs and the weak region(s) for enzymic attack.   Trials to identify the site by tryptic digestion on crystals have not been successful.   The electron micrographs of the larger component of tryptic digestion have illustrated the existence of molecules shorter than monomers, indicating the shape of fragments.

As for the end structure of the molecule, some facts must be pointed out.   First, the filaments in both the 200 Å network and the 400 Å square lattice seem to be smooth without any knot that could be recognized if molecules are overlapped at the connecting region.   The dots to indicate the repeating structure also have no disturbance at that region.   Therefore, the molecules appear to contact each other end-to-end.   In addition, the forces acting between both ends to hold the lattice structure must not be so strong, because the edge structures show the breaking of noncovalent bonds between molecules.   If a molecule, therefore, has smooth ends which contact alternately to form a polymer, the forces seem to be too weak.

From these results, one may postulate as one of the plausible models for the end structure that a small portion at both ends would remain a single peptide that makes a coiled-coil structure when bound to each other.   Such a binding mode could become more favorable by further side-by-side attachment of other molecules, thus making up not only end-to-end but also side-by-side aggregation to form a polymer, that is, the intermolecular interactions would stabilize the end-to-end binding of the molecules.   To ascertain the above model, studies on the nature of the interaction and the details of the structure are required.

LOCATION OF TROPOMYOSIN IN MYOFIBRIL

There are some experiments to identify the location of tropomyosin in myofibril (39, 49, 50).   The most plausible location is in the thin filaments that consist mainly of F-actin.   The identification of troponin in thin filaments is made by the attachment of anti-troponin to the filaments (51).   Since the binding of tropomyosin to actin and troponin is relatively strong, tropomyosin together with F-actin and troponin may constitute thin filaments.   Another location for tropomyosin might be in the Z-band (39), whose structure is a square lattice of 200 Å separation, quite similar to the one in the tropomyosin crystal.   However, the existence of tropomyosin molecules in the Z-band has not been confirmed so far.   As far as the resemblance in morphology of crystals and myo-fibrils is concerned, a hexagonal structure observed near the M-line is also observed in tropomyosin crystals.   To determine whether tropomyosin is located in places other than the thin filament, more extensive studies are needed.

SUMMARY

Tropomyosin, one of the major components of muscle proteins, has a role in the control of muscle contraction. The protein extracted from minced muscle with a concentrated salt solution is polymerizable at a low ionic strength. Physicochemical studies on polymerization-depolymerization properties indicate that molecules with a rod-like shape interact with each other not only end-to-end but side-by-side. The monomeric unit has a molecular weight of about 60,000 and a rod-like shape of 400 Å in length. A high helical content is consistent with the rod-shape. However biochemical studies using limited proteolysis with trypsin show that there are some loose region(s) which are easily attacked by the enzyme. Electron microscopic observations of tropomyosin crystals give information on the molecular structure which seems to be a coiled-coil of two α-helices. Binding of other proteins, especially troponin to tropomyosin, can be seen on the crystal. Finally, a possible structure of the tropomyosin molecule is discussed.

REFERENCES

1 K. Bailey, *Nature*, **157**, 368 (1946).
2 K. Bailey, *Biochem. J.*, **43**, 271 (1948).
3 J. C. Ruegg, *Proc. Roy. Soc.*, (*London*), *Ser B*, **154**, 209 (1961).
4 S. Ebashi, *Nature*, **200**, 1010 (1963).
5 S. Ebashi and A. Kodama, *J. Biochem.*, **58**, 107 (1965).
6 S. Ebashi, A. Kodama and F. Ebashi, *J. Biochem.*, **64**, 465 (1968).
7 D. R. Kominz, F. Saad and K. Laki, *Nature*, **179**, 206 (1957).
8 T. C. Tsao and K. Bailey, *Biochim. Biophys. Acta*, **11**, 102 (1953).
9 A. Szent-Györgyi and B. Kaminer, *Proc. Natl. Acad. Sci. U.S.*, **50**, 1033 (1963).
10 S. Higashi and T. Ooi, *J. Mol. Biol.*, **34**, 699 (1968).
11 S. Fujime-Higashi and T. Ooi, *J. Microscopie*, **8**, 535 (1969).
12 S. Iida and T. Ooi, *Arch. Biochem. Biophys.*, **121**, 526 (1967).
13 T. Ooi, K. Mihashi and H. Kobayashi, *Arch. Biochem. Biophys.*, **98**, 1 (1962).
14 A. Sato and K. Mihashi, personal communication.
15 H. Asai, *J. Biochem.*, **50**, 182 (1961).

16 K. Maruyama, *Sci. Papers Coll. Gen. Edu. Univ. Tokyo*, **9**, 147, (1959).

17 T. C. Tsao, K. Bailey and C. S. Adair, *J. Biochem.*, **49**, 27 (1951).

18 S. Iida and N. Imai, *J. Phys. Chem.*, **73**, 75 (1969).

19 W. Drabikowski and E. Nowak, *Acta Biochim. Polon.*, **12**, 61 (1965).

20 H. Mueller, *Biochem. Zeit.*, **345**, 300 (1966).

21 C. M. Kay and K. Bailey, *Biochim. Biophys. Acta*, **40**, 149 (1960).

22 A. Holtzer, R. Clark and S. Lowey, *Biochem.*, **4**, 2401 (1965).

23 K. Imahori and P. Doty, *Advan. in Prot. Chem.*, **16**, 401 (1961).

24 F. H. C. Crick, *Nature*, **170**, 882 (1952).

25 C. Peng, T. Kung, L. Hsiung and T. Tsao, *Scientia Sinica*, **14**, 219 (1965).

26 E. F. Woods, *J. Biol. Chem.*, **242**, 2859 (1967).

27 C. E. Bodwell, *Arch. Biochem. Biophys.*, **122**, 246 (1967).

28 J. Olander, M. F. Emerson and A. Holtzer, *J. Am. Chem. Soc.*, **89**, 3058 (1967).

29 K. Weber and M. Osborn, *J. Biol. Chem.*, **244**, 4406 (1969).

30 R. H. Locker, *Biochim. Biophys. Acta*, **14**, 533 (1954).

31 D. R. Kominz, F. Saad, J. A. Gladner and K. Laki, *Arch. Biochem. Biophys.*, **70**, 16 (1957).

32 F. Saad and D. R. Kominz, *Arch. Biochem. Biophys.*, **92**, 541 (1961).

33 R. E. Alving, E. Moczar and K. Laki, *Biochem. Biophys. Res. Commun.*, **23**, 540 (1966).

34 T. Ooi, *Biochem.*, **6**, 2433 (1967).

35 M. Jen, T. Hsu and T. Tsao, *Scientia Sinica*, **14**, 81 (1965).

36 Y. Tsou, Z. Lu and T. Tsao, *Scientia Sinica*, **15**, 83 (1966).

37 M. F. Perutz, *Europ. J. Biochem.*, **8**, 455 (1969).

38 J. Hanson and J. Lowy, *J. Mol. Biol.*, **6**, 46 (1963).

39 H. E. Huxley, *J. Mol. Biol.*, **7**, 281 (1963).

40 A. J. Hodge, *Rev. Mod. Phys.*, **31**, 409 (1959).

41 C. Cohen and W. Longley, *Science*, **152**, 794 (1966).

42 D. L. D. Casper, C. Cohen and W. Longley, *J. Mol. Biol.*, **41**, 87 (1969).

43 C. Cohen and K. C. Homes, *J. Mol. Biol.*, **6**, 423 (1963).

44 K. Laki, K. Maruyama and D. R. Kominz, *Arch. Biochem. Biophys.*, **98**, 323 (1962).

45 W. Drabikowski, Y. Nonomura and K. Maruyama, *J. Biochem.*, **63**, 761 (1968).

46 Y. Nonomura, W. Drabikowski and S. Ebashi, *J. Biochem.*, **64**, 419 (1968).

47 D. A. D. Parry and E. Suzuki, *Biopolymer*, **7**, 189 (1969).

48 D. A. D. Parry and E. Suzuki, *Biopolymer*, **7**, 199 (1969).

49 S. V. Perry, *Physiol. Rev.*, **36**, 1 (1956).

50 M. Endo, Y. Nonomura, T. Masaki, I. Ohtsuki and S. Ebashi, *J. Biochem.*, **60**, 605 (1966).

51 I. Ohtsuki, T. Masaki, Y. Nonomura and S. Ebashi, *J. Biochem.*, **61**, 817 (1967).

52 Y. Tawada and T. Ooi, unpublished results.

Received for publication, June 10, 1970.

Advan. in Biophys., Vol. 2, pp. 155–200 (1971)

STOCHASTIC THEORY OF REACTION KINETICS

EI TERAMOTO, NANAKO SHIGESADA,
HISAO NAKAJIMA and KOUJI SATO

Department of Biophysics, Kyoto University, Kyoto, Japan

The biological function of a living system is a highly organized assembly of chemical reactions, in which various elementary reaction processes mutually have space and time correlations.   In order to analyze the time development of this kind of process, we need a generalized systematic formulation for chemical reaction kinetics.

In a purely physical sense, the chemical reaction process is nothing but the time development of the change of the microscopic state of a certain system.   It may consist of the motion of constituent molecules or parts of the system, transitions of their quantum states, so-called chemical reactions such as molecular binding or dissociation and so on. However, for a given individual problem, we must choose a plausible way of specifying the microscopic states of the system according to the nature of the problem.

In this article, we shall develop a generalized stochastic formula for chemical reaction processes.   In Part I, we shall present a general formula that can be used to describe formally the time development of the state of reaction systems and compare it with the usual technique of the stochastic theory of chemical kinetics.

In Part II, to illustrate an application of this technique, the diffusion-

155

controlled reaction kinetics is treated as a many-body problem and the effects of three- and four-body correlations are discussed, using the cluster expansion method.

Finally, in Part III, we shall present the results of the computer simulation of the lattice model of diffusion-controlled bimolecular reaction processes.

I. GENERAL FORMULA

1. *Introduction*

The stochastic approach to the problem of chemical kinetics was first presented by Kramers (*12*) who investigated the rate of transition over a barrier from state A to B, using the Smoluchowski's equation. This kind of problem was generalized later by Shuler *et al.* (*11, 22, 27*), who applied a discrete random walk model to the various states of the molecules.

The stochastic formula, which describes the time development of the reaction system such as A→B or A+B→C, was first developed by Rényi (*25*). He investigated differential-difference equations for the probability of finding, for instance, n A molecules in the system at time t. Then one can calculate the fluctuations as well as the averaged values and also examine the validity of the deterministic law given by the law of mass action. This sort of problem was extended by Bartholomay (*2*), Ishida (*9*), McQuarrie (*16, 17*) and others (*6, 7, 30*). In these stochastic equations, a constant reaction rate is usually assumed.

On the other hand, many authors (*5, 19–21, 31, 34*) have developed theoretical investigations on the diffusion-controlled reaction process since Smoluchowski's fundamental work, which was originally developed to explain the process of the coagulation of colloids, appeared (*29*). Many of these works were devoted to deriving formulae for collision frequency or reaction rate as a function of time and friction constant on the basis of Smoluchowski's equation. Assuming the independence of binary collisions, they solved essentially two-body problems under various boundary and initial conditions.

Furthermore, there are many interesting stochastic problems concerning more complicated systems such as chemical reactions related to polymers or biological macromolecules. The variety of problems is quite extensive and a special stochastic approach was proposed for each problem.   The wide range of applications of the stochastic theory in regard to chemical reactions is discussed in review articles by Mc-Quarrie (*18*), Montroll (*23*) and Weiss (*33*).

Here the chemical reaction processes are formulated from a more fundamental point of view of the stochastic processes.   As a basic probability function, we first introduced the probability of the first reaction time.   It will be shown that the time development of the ensemble of the reaction system can be expressed in terms of this probability function.   When one considers a stationary reaction process in which the time dependence of the probability of the first reaction time is given only through the time interval, the problem becomes rather simple mathematically and the formal solution is obtained by the use of the Laplace transformation.   Furthermore, if we assume the constant reaction rate, the formula is reduced to the master equation.

The state of the reaction system can be specified, in principle, by microscopic quantities.   However, in actual experimental analysis, macroscopic quantities such as the density of molecules are used. Therefore, if one starts from the analysis of the time development of the microscopic state of the system, it is necessary to reduce it to the formula which describes the change of the macroscopic state.   This situation is quite similar to the relation between the statistical mechanics and the thermodynamical laws.

2.   *Definitions and mathematical formula*

We are going to consider a system which is capable of assuming discrete states x_i ($i=1, 2, \ldots, N$), where x_i may be the value of some quantity, or the values of a set of quantities, of this system.   Because of some kinds of chemical reactions which occur in the system, the system itself undergoes successive transitions between these possible states.

Here we shall introduce the probability of the first reaction time, that is, $T(x_j, t|x_i, t_0)$ dt ($j \neq i$) is the probability that, starting from state x_i at time t_0, the state of the system remains unchanged and no reaction occurs until time t and in the time interval from t to $t+$dt, a reaction occurs for the first time and the state changes from x_i to x_j.   The probability that, starting from state x_i, the first reaction occurs during the time interval $(t, t+$d$t)$ is given by

$$\tau(\bar{x}_i, t \mid x_i, t_0)dt = \sum_{\substack{j=1 \\ j \neq i}}^{N} T(x_j, t \mid x_i, t_0)dt , \qquad (2.1)$$

where $\bar{x}_i$ denotes all possible states, excluding the initial state x_i.

Furthermore, we shall introduce the survival probability of state x_i, $P(x_i, t \mid x_i, t_0)$, which is the probability that, starting from state x_i at time t_0, the system remains unchanged and no reaction occurs until time t. Then the normalization condition can be expressed by the equation:

$$\int_{t_0}^{t} \tau(\bar{x}_i, t' \mid x_i, t_0)dt' + P(x_i, t \mid x_i, t_0) = 1 . \qquad (2.2)$$

Using these two kinds of probabilities, the probability of the first reaction time and the survival probability, we can express the probability that, starting from the initial state x_i, the system is found in state x_j at time t after successive transitions of n times, as follows:

$$U_n(x_j, t \mid x_i, 0) = \sum_{i_1,\ldots,i_{n-1}} \int_0^t dt_n \int_0^{t_n} dt_{n-1} \ldots \int_0^{t_2} dt_1 P(x_j, t \mid x_j, t_n)$$

$$\times T(x_j, t_n \mid x_{i_{n-1}}, t_{n-1}) T(x_{i_{n-1}}, t_{n-1} \mid x_{i_{n-2}}, t_{n-2}) \ldots$$

$$\times T(x_{i_1}, t_1 \mid x_i, 0) , \qquad n \geq 1 , \qquad (2.3)$$

where the summations are taken over all possible states under the conditions $x_{i_k} \neq x_{i_{k+1}}$ ($k=0, 1, \ldots, n-1$ and $i_0=i, i_n=j$).

It can be easily seen that the probability of finding the system in state x_j at time t is given by

$$W(x_j, t \mid x_i, 0) = \sum_{n=1}^{\infty} U_n(x_j, t \mid x_i, 0) + \delta_{ij} P(x_i, t \mid x_i, 0) . \qquad (2.4)$$

In a case where the initial and final states are the same, the second term of Eq. 2.4 is necessary.

The function defined by

$$s_n(t \mid x_i, 0) = \sum_{j=1}^{N} U_n(x_j, t \mid x_i, 0) , \qquad n \geq 1 , \qquad (2.5)$$

gives the probability that the system, starting from state x_i, suffers successive transitions of n times until time t. Then the probability that a reaction occurs at least once until time t is given by

$$S(t \mid x_i, 0) = \sum_{n=1}^{\infty} s_n(t \mid x_i, 0) \,. \tag{2.6}$$

If we keep in mind the definition of the probability of the first reaction time, it is easy to see that the identity:

$$S(t \mid x_i, 0) = \int_0^t \tau(\bar{x}_i, t' \mid x_i, 0) \mathrm{d}t' \tag{2.7}$$

holds. Actually, if we take the sum of the final state x_j in Eq. 2.4 and use Eqs. 2.5, 2.6, 2.7 and 2.2, we have the normalization condition of $W(x_j, t \mid x_i, 0)$ as

$$
\begin{aligned}
\sum_{j=1}^{N} W(x_j, t \mid x_i, 0) &= \sum_{j=1}^{N} \sum_{n=1}^{\infty} U_n(x_j, t \mid x_i, 0) + P(x_i, t \mid x_i, 0) \\
&= S(t \mid x_i, 0) + P(x_i, t \mid x_i, 0) \\
&= \int_0^t \tau(\bar{x}_i, t' \mid x_i, 0) \mathrm{d}t + P(x_i, t \mid x_i, 0) \\
&= 1 \,.
\end{aligned}
\tag{2.8}
$$

When we do not have definite information about the initial state and only the initial probability distribution $\rho(x_i)$ can be attributed, the probability of finding the system in state x_j at time t after successive transition of n times is given by

$$U_n(x_j, t) = \sum_{i=1}^{N} U_n(x_j, t \mid x_i, 0) \rho(x_i) \,. \tag{2.9}$$

Using Eq. 2.4, the probability of finding the system in state x_j at time t is obtained as

$$W(x_j, t) = \sum_{n=1}^{\infty} U_n(x_j, t) + P(x_j, t \mid x_j, 0) \rho(x_j) \,, \tag{2.10}$$

and the expectation value of a quantity, which is a function of state $f(x_j)$, is given by

$$\langle f \rangle = \sum_{j=1}^{N} f(x_j) W(x_j, t) \,. \tag{2.11}$$

3. Reaction rate function and master equation

Let us consider the case of stationary reaction processes in which the time dependence of the probability of the first reaction time is only given through the time interval $t-t_0$, that is

$$T(x_j, t \mid x_i, t_0)\mathrm{d}t = T(x_j \mid x_i;\ t-t_0)\mathrm{d}t\,, \qquad (i, j = 1, 2, \ldots, N;\ i \neq j)\,. \quad (3.1)$$

Then it is obvious that the survival probability $P(x_i, t \mid x_i, t_0)$ also depends only on $t-t_0$ because of Eq. 2.2, so we shall denote it by $P(x_i, t-t_0)$.

For the stationary reaction process, we shall define the rate function of the reaction from the state x_i to x_j, $k(x_j \mid x_i;\ t-t_0)$, by the equation:

$$k(x_j \mid x_i;\ t-t_0)\mathrm{d}t = \frac{T(x_j \mid x_i;\ t-t_0)}{P(x_i;\ t-t_0)}\mathrm{d}t\,, \qquad (i, j = 1, 2, \ldots, N;\ i \neq j)\,, \quad (3.2)$$

and the total rate function by

$$K(\bar{x}_i \mid x_i;\ t-t_0)\mathrm{d}t = \frac{\tau(\bar{x}_i \mid x_i;\ t-t_0)}{P(x_i, t-t_0)}\mathrm{d}t$$

$$= -\frac{\mathrm{d}}{\mathrm{d}t} \log P(x_i, t-t_0)\mathrm{d}t\,, \qquad\qquad (3.3)$$

where for the second equality we used a relation which can be easily derived from Eq. 2.2, that is

$$\frac{\mathrm{d}}{\mathrm{d}t}P(x_i, t-t_0) = -\sum_{j=1}^{N} T(x_j \mid x_i;\ t-t_0)\,. \qquad (3.4)$$

In the case of stationary reaction processes, the multiple integral in Eq. 2.3 can be calculated, in principle, using the Laplace transformation. If we denote the Laplace transform of the function of type $F(x, t)$ by

$$F(x, s) = \int_0^\infty F(x, t)e^{-st}\mathrm{d}t\,, \qquad (3.5)$$

the Laplace transform of Eq. 2.3 can be written in the form:

$$U_n(x_j, s) = \sum_{i, i_1, \ldots, i_{n-1}} P(x_j, s)T(x_j \mid x_{i_{n-1}};\ s)\ldots T(x_{i_1} \mid x_i;\ s)\rho(x_i)$$

$$= \sum_{i, i_1, \ldots, i_{n-1}} P(x_j, s) \prod_{k=1}^{n} T(x_{i_k} \mid x_{i_{k-1}};\ s)\rho(x_i)\,. \qquad (i_0 = i,\ i_n = j) \quad (3.6)$$

Thus considering the vectors defined by

$$\mathbf{U}_n = \begin{pmatrix} U_n(x_1, s) \\ U_n(x_2, s) \\ \vdots \\ U_n(x_N, s) \end{pmatrix}, \qquad \boldsymbol{\rho} = \begin{pmatrix} \rho(x_1) \\ \rho(x_2) \\ \vdots \\ \rho(x_N) \end{pmatrix},$$

(3.7)

and matrices:

$$\mathbf{P} = \begin{pmatrix} P(x_1, s) & 0 & 0 & \cdots & 0 \\ 0 & P(x_2, s) & 0 & \cdots & 0 \\ & \cdots\cdots\cdots & & & \\ & \cdots\cdots\cdots & & & \\ 0 & 0 & 0 & \cdots & P(x_N, s) \end{pmatrix},$$

$$\mathbf{T} = \begin{pmatrix} 0 & T(x_1 \,|\, x_2\,;\, s) & \cdots & T(x_1 \,|\, x_N\,;\, s) \\ T(x_2 \,|\, x_1\,;\, s) & 0 & & T(x_2 \,|\, x_N\,;\, s) \\ & \cdots\cdots\cdots & & \\ & \cdots\cdots\cdots & & \\ T(x_N \,|\, x_1\,;\, s) & T(x_N \,|\, x_2\,;\, s) & \cdots & 0 \end{pmatrix},$$

(3.8)

Eq. 3.6 can be expressed as

$$\mathbf{U}_n = \mathbf{P}\mathbf{T}^n \boldsymbol{\rho} .$$

(3.9)

Then, the Laplace transform of the probability $W(x_j, t)$, Eq. (2.10), is given by the matrix form:

$$\mathbf{W} = \mathbf{P}[1 + \sum_{n=1}^{\infty} \mathbf{T}^n]\boldsymbol{\rho} = \mathbf{P}[1-\mathbf{T}]^{-1}\boldsymbol{\rho} .$$

(3.10)

This expression looks very simple, but the actual problems are usually too complicated to be solved by analytical calculation.

There is another approach applicable to the reaction processes which satisfies the following condition for the rate functions. If we assume that the rate functions in Eq. 3.2 are constants with respect to time:

$$k(x_j \,|\, x_i\,;\, t-t_0)dt = k(x_j \,|\, x_i)dt ,$$

(3.11)

then it can be shown that our formula is reduced to a well-known fundamental equation, that is, the master equation for chemical kinetics. In order to see it, we shall consider the time derivative of the probability $W(x_j, t)$. From Eq. 2.10 we have

$$\frac{\mathrm{d}}{\mathrm{d}t}W(x_j, t) = \sum_{n=1}^{\infty} \frac{\mathrm{d}}{\mathrm{d}t}U_n(x_j, t) + \frac{\mathrm{d}}{\mathrm{d}t}P(x_j, t)\rho(x_j). \qquad (3.12)$$

Using Eqs. 2.3 and 2.9, we obtain

$$
\begin{aligned}
\frac{\mathrm{d}}{\mathrm{d}t}U_n(x_j, t) = {}& \sum_{i,i_1,\dots,i_{n-1}} \int_0^t \mathrm{d}t_{n-1}\dots \int_0^{t_2}\mathrm{d}t_1\, T(x_j\,|\,x_{i_{n-1}};\ t-t_{n-1}) \\
& \times T(x_{i_{n-1}}\,|\,x_{i_{n-2}};\ t_{n-1}-t_{n-2})\dots T(x_{i_1}\,|\,x_i;\ t_1)\rho(x_i) \\
& - \sum_{i,i_1,\dots,i_{n-1}} \int_0^t \mathrm{d}t_n \int_0^{t_n}\mathrm{d}t_{n-1}\dots \int_0^{t_2}\mathrm{d}t_1 \sum_k T(x_k\,|\,x_j;\ t-t_n) \\
& \times T(x_j\,|\,x_{i_{n-1}};\ t_n-t_{n-1})\dots T(x_{i_1}\,|\,x_i;\ t_1)\rho(x_i), \qquad n\geqq 2,\ (3.13)
\end{aligned}
$$

where we used the relation $P(x_j, 0)=1$ and Eq. 3.4. In Eq. 3.13 if we use the definition of the rate function (3.2) and also the assumption (3.11), we have

$$
\begin{aligned}
\frac{\mathrm{d}}{\mathrm{d}t}U_n(x_j;\ t) = {}& \sum_{i_{n-1}} k(x_j\,|\,x_{i_{n-1}}) \sum_{i,i_1,\dots,i_{n-2}} \int_0^t \mathrm{d}t_{n-1} \int_0^{t_{n-1}}\mathrm{d}t_{n-2}\dots \int_0^{t_2}\mathrm{d}t_1 \\
& \times P(x_{i_{n-1}}, t-t_{n-1})T(x_{i_{n-1}}\,|\,x_{i_{n-2}};\ t_{n-1}-t_{n-2})\dots \\
& \times T(x_{i_1}\,|\,x_i;\ t_1)\rho(x_i) \\
& - \sum_k k(x_k\,|\,x_j) \sum_{i,i_1,\dots,i_{n-1}} \int_0^t \mathrm{d}t_n \int_0^{t_n}\mathrm{d}t_{n-1}\dots \int_0^{t_2}\mathrm{d}t_1 \\
& \times P(x_j, t-t_n)T(x_j\,|\,x_{i_{n-1}};\ t_n-t_{n-1})\dots \\
& \times T(x_{i_1}\,|\,x_i;\ t_1)\rho(x_i), \qquad n\geqq 2,
\end{aligned}
\qquad (3.14)
$$

and using Eq. 2.3, we can get

$$\frac{\mathrm{d}}{\mathrm{d}t}U_n(x_j, t) = \sum_k [k(x_j\,|\,x_k)U_{n-1}(x_k, t) - k(x_k\,|\,x_j)U_n(x_j, t)], \qquad n\geqq 2.\ (3.15)$$

For $n=1$ we can easily obtain the relation:

$$\frac{\mathrm{d}}{\mathrm{d}t}U_1(x_j, t) = \sum_i k(x_j\,|\,x_i)P(x_i, t)\rho(x_i) - \sum_k k(x_k\,|\,x_j)U_1(x_j, t). \qquad (3.16)$$

Substituting Eqs. 3.15 and 3.16 into Eq. 3.12 we obtain finally

$$\frac{\mathrm{d}}{\mathrm{d}t}W(x_j, t) = \sum_k [k(x_j\,|\,x_k)W(x_k, t) - k(x_k\,|\,x_j)W(x_j, t)]. \qquad (3.17)$$

This is the master equation for chemical reaction kinetics.

4. *The specification of state*

According to a purely physical interpretation, the chemical reaction processes are nothing but the time development of the change of the state of a macroscopic system. Therefore, if the state is specified by the microscopic variables x, where, as already mentioned, x generally denotes a set of variables, the transition from state x_i to x_j represents an elementary physical process or, for instance, the quantum mechanical transition. Then the rate function $k(x_j|x_i;\ t-t_0)$ would be evaluated by quantum mechanical law; and in many cases, the rate function for these kinds of elementary processes can be assumed to give a constant transition probability $k(x_j|x_i)$ under some suitable assumptions.

However, in actual experiments of chemical reaction processes, we usually measure the macroscopic quantity. If the state is specified by such a macroscopic quantity x, even when the constant transition probabilities can be assumed for elementary processes, the rate functions of the macroscopic state, generally, should be considered as functions of time.

For example, consider the bimolecular reaction processes of type $A+B \rightleftharpoons C$ in a liquid medium or in a gas phase. In this case the macroscopic quantities which are actually measured to analyze this reaction kinetics are the densities of the molecules. The simplest formula to use in analyzing this kind of process is the law of mass action:

$$\frac{\mathrm{d}}{\mathrm{d}t}C_\mathrm{A} = -\kappa_1 C_\mathrm{A} C_\mathrm{B} + \kappa_2 C_\mathrm{C}\,,$$

where C_A, C_B and C_C are the densities of A, B and C molecules, respectively.

However, when the system is under microscopic inspection, it is evident that the elementary processes of the system are rather complicated, namely, every molecule always changes its position by thermal motion and when two molecules, A and B, come into the reaction range, there is some probability that a chemical reaction will occur. Therefore, the microscopic state of this system is represented by the coordinates of the positions of all the molecules. The velocities of all the molecules are also necessary, especially in the case of reaction processes in the gas phase.

As already mentioned in the introduction, these problems were investigated by many authors. However, we want to reexamine the problems, using our systematic formulas developed above. As a formal procedure, starting from the expression which describes the change of the microscopic state, we must reduce it to the formula which describes directly the change of the macroscopic state.

In the following sections, as a special example, we shall treat bimolecular reaction processes in a liquid medium.

II. BIMOLECULAR REACTIONS IN A LIQUID MEDIUM

5. *Diffusion-controlled reaction kinetics*

In the study of reaction kinetics in solution, diffusion-controlled reaction provides an interesting model system for the theoretical investigation of the reaction process from the stochastic approach. The first important contribution to this problem was made by Smoluchowski in his mathematical treatment of the coagulation of colloidal particles (29). The equation presented by him has been extended to include ordinary bimolecular reaction kinetics by many authors. Reactions which involve an excitation transfer, such as the quenching of fluorescence (5, 24, 31, 34) and radio chemical reaction (20) are typical examples that have been examined. It has also been applied to the catalytic process (32), including enzymic reactions (7). From these studies, many interesting results have been obtained about the collision frequency and the reaction rate as a function of the time and friction constant.

However, these authors treated the reaction processes as essentially two-body problems, assuming the independence of collisions. If we consider the reaction process as a time dependent many-body problem, it becomes necessary to take into account the complicated space and time correlation. Although the effects of these correlations have been discussed by several authors, a systematic formula has not yet been established. The purpose of the remaining part of this chapter is to describe the calculation of diffusion-controlled reaction kinetics as a special application of the general formula presented above.

Before advancing to the calculation, we shall outline the theoretical approaches of previous authors and their important results to illuminate the later discussions. The principle of Smoluchowski's treatment is based on the assumption that any pair of colloidal particles will adhere to each other instantaneously on collision. It is also assumed that particles undergo random motion in a liquid medium which can be described by the diffusion equation and that the reaction rate is a function of the number of collisions per unit time.

The diffusion equation (Smoluchowski's equation) for the probability of finding two particles at a relative distance r at time t, $\rho(r, t)$, is written as

$$\frac{\partial}{\partial t}\,\rho(r, t) = D\nabla^2 \rho(r, t), \tag{5.1}$$

where D is the relative diffusion constant and ∇^2 is the Laplace operator. This equation is solved under the following conditions: 1) At the initial time, particle A at $r=0$ is surrounded by B particles of a uniform concentration ρ_0. 2) At a radius σ from the central particle A, the concentration of B particles is always zero as a result of the reaction and at a very distant point, it remains at the initial value ρ_0. Under the above conditions, we can solve Eq. 5.1. The rate of reaction is given by the flux of B particles into the sphere of radius σ around the A particle in the form:

$$4\pi D\sigma\left(1+\frac{\sigma}{\sqrt{\pi Dt}}\right)\rho_0. \tag{5.2}$$

It shows that the time dependent term contributes to the rate of reaction mainly in the time range $t<\sigma^2/\pi D$, which is important in describing the quenching of fluorescence.

Many authors applied Smoluchowski's treatment of the problem of diffusion limited bimolecular reactions. They solved diffusion equations with different sets of initial and boundary conditions. Among them, there are two representative examples of boundary conditions. They are 1) the Smoluchowski boundary condition:

$$\rho(\sigma) = 0,$$

which has already been mentioned and 2) the radiation boundary condition:

$$D\frac{\partial}{\partial r}\rho(r, t) = \beta\rho, \qquad r \to \sigma,$$

which means that only a certain fraction of the encounters of reactive particles leads to reactions and the rate of reaction which is equal to the flux of B particles through the sphere of radius σ is proportional to the concentration of B particles on the sphere. Further, Debye (*8*) and more recently Montroll (*21*) extended Smoluchowski's treatment to the case of the quenching of fluorescence where both fluorescing and quenching particles are ions and the thermal diffusion process interferes with the ionic interactions. Recently, Monchick, Magee and Samuel have studied radiation chemistry as a many-particle reaction process taking into account the correlation of the density function of the individual particles. For the calculation of the correlation function, they used a superposition approximation.

In order to evaluate the many-body effect on the reaction rate, we reexamined the problems of diffusion-controlled reaction, using the Smoluchowski and radiation boundary conditions with our systematic formula developed above.

6. *Mathematical formula*

Let us consider a solution containing two kinds of particles, A and B, as solutes in a liquid medium of volume V. The particles are free to diffuse with diffusion constants D_A and D_B and an irreversible reaction between the A and B particles of type A+B→C occurs when any pair of A and B particles approaches within a certain range. The probability that a reaction between a pair of A and B particles at a relative distance r, during the time interval dt, is formally expressed by $k(r)dt$.

In the following discussion, it is assumed that the direct intermolecular interactions, except for the reaction between A and B particles, which may modify the random motion, can be neglected. As an initial ensemble, we are going to consider an ensemble of the systems in which N particles of type A and $M(M \geq N)$ particles of type B are distributed in a medium of volume V and no particles of type C are formed at $t=0$. At $t>0$, these particles begin thermal motions. When the two species of particles approach each other within the reaction range, a reaction of type A+B→C will take place. We shall denote by $U(N-n|N;\ t)$ the probability that at time t, n C particles have already formed, so that $N-n$ A and $M-n$ B particles remain unreacted. Using this probability function, the average density of A particles at time t is given by

$$C_A(t) = \frac{1}{V} \sum_{n=0}^{N} (N-n)U(N-n\,|\,N;\ t). \tag{6.1}$$

In order to take into account the time and space correlations of the reaction processes of our many-particle system, we shall introduce the variables $X=(x_1, x_2, \ldots, x_\mu)$ as a set of coordinates of μ A particles and denote the position vector of the i-th A particle by x_i. Similarly, $Y=(y_1, y_2, \ldots, y_\nu)$ is defined for B particles.

Now the probability of the first reaction time $T_{\mu,\nu}{}^{i,j}(X, Y; t|X_0, Y_0; t_0)$ $dXdYdt$ is defined as follows: the probability that until time t no reaction has occurred and $\mu+\nu$ particles of types A and B, which have started from the positions (X_0, Y_0), are found in the volume element $dXdY$ at the positions (X, Y) at time t and in the time interval from t to $t+dt$ the first reaction occurs between the i-th A and j-th B particle at the positions x_i and y_j, respectively.

Apparently the probability of the first reaction time can be expressed in the form:

$$T_{\mu,\nu}{}^{i,j}(X, Y; t \mid X_0, Y_0; t_0)dXdYdt$$
$$= k(x_i, y_j)P_{\mu,\nu}(X, Y; t \mid X_0, Y_0; t_0)dXdYdt, \qquad (6.2)$$

where $k(x_i, y_j)=k(|x_i-y_j|)$ is the intrinsic reaction rate defined above and $P_{\mu,\nu}(X, Y; t|X_0, Y_0; t_0)$ $dXdY$ is the survival probability that $\mu+\nu$ particles of types A and B, which started from the positions (X_0, Y_0) at time t_0, have not reacted yet and can be found in the volume element $dXdY$ at the positions (X, Y) at time t.

Then the probability that n C particles are found at time t, $U(N-n|N; t)$ can be constructed by summing up the probabilities of each individual process which leads to the formation of n C particles at time t. Namely, it can be written in the form:

$$U(N-n \mid N; t) = \sum_{\{(i_k, j_k)\}} \int_0^t dt_n \int_0^{t_n} dt_{n-1} \ldots \int_0^{t_2} dt_1 \int dXdY$$

$$\times \int dX_n dY_n \ldots \int dX_0 dY_0$$

$$\times P_{N-n, M-n}(X/[i_1 \ldots i_n], Y/[j_1 \ldots j_n]; t$$

$$\mid X_n/[i_1 \ldots i_n], Y/[j_1 \ldots j_n]; t_n)$$

$$\times T_{N-n+1, M-n+1}{}^{i_n, j_n}(X_n/[i_1 \ldots i_{n-1}], Y_n/[j_1 \ldots j_{n-1}]; t_n$$

$$\mid X_{n-1}/[i_1 \ldots i_{n-1}], Y_{n-1}/[j_1 \ldots j_{n-1}]; t_{n-1})$$

$$\times T_{N-n+2, M-n+2}{}^{i_{n-1}, j_{n-1}}(X_{n-1}/[i_1 \ldots i_{n-2}],$$

$$Y_{n-1}/[j_1 \ldots j_{n-2}]; t_{n-1} \mid X_{n-2}/[i_1 \ldots i_{n-2}],$$

$$Y_{n-2}/[j_1{\cdots}j_{n-2}]\,;\ t_{n-2})$$

$$\times\cdots$$

$$\times T_{N,M}{}^{i_1,j_1}(X_1,\,Y_1\,;\ t_1\,|\,X_0,\,Y_0\,;\ t_0)\rho_0(X_0,\,Y_0)\,,\quad (6.3)$$

where the notations $X_k/[i_1,\,i_2,\,\ldots,\,i_s]$ and $Y_k/[j_1,\,j_2,\,\ldots,\,j_s]$ mean the coordinates of $N-s$ A and $M-s$ B particles at time t_k, excluding the coordinates of $i_1,\,\ldots,\,i_s$-th A particles and $j_1,\,\ldots,\,j_s$-th B particles, which have already reacted. The summation is taken over all possible sequences of n different AB pairs $(i_1,\,j_1)$, $(i_2,\,j_2)$, $\ldots$, $(i_n,\,j_n)$, where $i_k=1,\,2,\,\ldots,$ N, $j_k=1,\,2,\,\ldots,\,M$ and $i_k\neq i_l$, $j_k\neq j_l$ for $k\neq l$. $\rho(X_0,\,Y_0)$ is the distribution function of particles of the initial ensemble.

Expression 6.3 gives detailed information on the reaction process of our system, including the space and time correlation in complete form. However, the calculation of Eq. 6.3 is virtually impossible. It is necessary to reduce the equation to a simpler form by using suitable approximations.

7. *Reduced formulas*

If, at the initial time, the particles A and B are uniformly distributed in the medium of volume V, then the chance of a reaction of the AB pair can be expected to be uniform in the space. In this case, we can reduce our formula to a simpler form by assuming that, at each instant t just after the k-th reaction has occurred, the remaining unreacted particles are distributed uniformly. Under this assumption, the multiple integration with respect to space coordinates in Eq. 6.3 can be performed independently; and we have

$$U(N-n\,|\,N\,;\ t)=\int_0^t \mathrm{d}t_n \int_0^{t_n} \mathrm{d}t_{n-1}\cdots \int_0^{t_2} \mathrm{d}t_1 P(N-n,\,t-t_n)$$

$$\times T(N-n\,|\,N-n+1\,;\ t_n-t_{n-1})$$

$$\times T(N-n+1\,|\,N-n+2\,;\ t_{n-1}-t_{n-2})$$

$$\times\cdots T(N-1\,|\,N\,;\ t_1)\,,\qquad n\geqq 1\,,\qquad (7.1)$$

where $(TN-k-1|N-k;\ t)$ is the reduced probability of the first reaction time that at the initial time the $N-k$ A and $M-k$ B particles are uniformly distributed in the medium of volume V, and in the time interval from t to $t+\mathrm{d}t$ the first reaction occurs within one of the AB pairs. $P(N-k,\,t)$ is the reduced survival probability that $N-k$ A and

$M-k$ B particles with uniform initial distribution remain unreacted until time t. Obviously, these probability functions are given by

$$P(N-k, t) = \int P_{N-k,M-k}(X, Y; t \mid X_0, Y_0; 0)$$
$$\times \rho(X_0, Y_0)\mathrm{d}X_0\mathrm{d}Y_0\mathrm{d}X\mathrm{d}Y, \qquad (7.2)$$

$$T(N-k-1 \mid N-k; t) = \sum_{i,j} \int k(x_i, y_j) P_{N-k,M-k}(X, Y; t \mid X_0, Y_0; 0)$$
$$\times \rho(X_0, Y_0)\mathrm{d}X_0\mathrm{d}Y_0\mathrm{d}X\mathrm{d}Y, \qquad (7.3)$$

where X_0 and X stand for the coordinates of the $N-k$ A particles, Y_0 and Y, the coordinates of the $M-k$ B particles, and $\rho(X_0, Y_0)$ denotes the uniform distribution function, that is, $\rho = V^{-N-M+2k}$. Because of the normalization condition of the probabilities, it is clear that

$$\int_0^t T(N-k-1 \mid N-k; t')\mathrm{d}t' + P(N-k, t) = 1, \qquad (7.4)$$

or in a differential form:

$$T(N-k-1 \mid N-k; t) = -\mathrm{d}P(N-k, t)/\mathrm{d}t. \qquad (7.5)$$

Thus, if we can obtain an explicit form of the distribution function $P_{N-k, M-k}(X, Y; t \mid X_0, Y_0; 0)$, the probability of the first reaction time $T(N-k-1 \mid N-k; t)$ is given by calculating Eqs. 7.2 and 7.5. Then, using Eqs. 6.1 and 7.1, we can evaluate the average density of A particles $C_A(t)$ as a function of time. Furthermore, we shall obtain the equation about the time derivative of $C_A(t)$. By differentiating $U(N-n \mid N; t)$ of Eq. 7.1 with respect to t, we have

$$\frac{\mathrm{d}}{\mathrm{d}t} U(N-n \mid N; t) = \int_0^t \mathrm{d}t_{n-1} \int_0^{t_{n-1}} \mathrm{d}t_{n-2} ... \int_0^{t_2} \mathrm{d}t_1 T(N-n \mid N-n+1; t-t_{n-1})$$
$$\times T(N-n+1 \mid N-n+2; t_{n-1}-t_{n-2}) ... T(N-1 \mid N; t_1)$$
$$- \int_0^t \mathrm{d}t_n \int_0^{t_n} \mathrm{d}t_{n-1} ... \int_0^{t_2} \mathrm{d}t_1 T(N-n-1 \mid N-n; t-t_n)$$
$$\times T(N-n \mid N-n+1; t_n-t_{n-1}) ... T(N-1 \mid N; t_1). \qquad (7.6)$$

Then from the definition of the average density of A particles in Eq. 6.1, we can obtain the following relation:

$$\frac{d}{dt} C_A(t) = \frac{1}{V} \sum_{n=0}^{N} (N-n) \frac{d}{dt} U(N-n \mid N; t)$$

$$= \frac{-1}{V} \sum_{n=1}^{N} \int_0^t dt_{n-1} \int_0^{t_{n-1}} dt_{n-2} \ldots \int_0^{t_2} dt_1 T(N-n \mid N-n+1; t-t_{n-1})$$

$$\times T(N-n+1 \mid N-n+2; t_{n-1}-t_{n-2}) \ldots T(N-1 \mid N; t_1). \tag{7.7}$$

In this equation, the multiple integral has the convolution form. If we can obtain the Laplace transform of $T(N-i-1|N-i; t)$, the calculation of Eq. 7.7 can be achieved by the following inverse Laplace transformation:

$$\frac{d}{dt} C_A(t) = \frac{-1}{V} \mathscr{L}^{-1} \left(\sum_{n=1}^{N} \prod_{i=0}^{n-1} T(N-i-1 \mid N-i; s); t \right), \tag{7.8}$$

where $T(N-i-1|N-i; s)$ is the Laplace transform of $T(N-i-1|N-i; t)$.

As a special case, let us consider the constant reaction rate. Namely, if the rate function is independent of t and given by

$$K_{n-k}(t) = \frac{T(N-k-1 \mid N-k; t)}{P(N-k, t)} = -\frac{d}{dt} \log P(N-k, t) = \kappa(N-k)(M-k),$$

as previously stated in Part I, our formula can be reduced to the master equation; and from Eq. 3.15 or 3.17, we have

$$\frac{d}{dt} U(N-n \mid N; t) = \kappa(N-n+1)(M-n+1) U(N-n+1 \mid N; t)$$

$$- \kappa(N-n)(M-n) U(N-n \mid N; t). \tag{7.9}$$

This equation has already been investigated by many authors and the exact solution obtained $(2, 9, 16, 17, 25)$. Using Eq. 7.9, we can easily derive the equation for the density of A particles as

$$\frac{d}{dt} C_A(t) = \frac{-\kappa}{V} \sum_{k=0}^{N} (N-k)(M-k) U(N-k \mid N; t)$$

$$\equiv -\frac{\kappa}{V} \langle (N-k)(M-k) \rangle . \tag{7.10}$$

Furthermore, if we assume that

$$\frac{1}{V^2}\langle(N-k)(M-k)\rangle \fallingdotseq \frac{1}{V^2}\langle N-k\rangle\langle M-k\rangle = C_A C_B,$$

Eq. 7.10 leads to the usual phenomenological kinetic equation of bimolecular reactions, the so-called law of mass action.

8. *Multi-dimensional diffusion equation with pair absorbing interaction*

In the last section, we derived a general formula to use in calculating the averaged density as a function of time. Thus, we found out that we had to calculate the distribution function, $P_{N,M}(X,\ Y;\ t|X_0,\ Y_0;\ t_0)$, first. The change of the probability distribution function P with time is attributable to the diffusive random motion of the particles and also to the reaction processes between the AB pairs.

In general, the microscopic state of this system should be specified by the coordinates and velocities of all the A, B and C particles. The diffusive motion of these particles is described in a phase space by the Fokker-Planck type equation (4). However, when the friction constant ζ is large, the Maxwellian velocity distribution will be established very quickly, namely, after the lapse of time of the order m/ζ, where m is the mass of the particle. After that time, the change of spatial distribution is characterized by the diffusion constant $D=kT/\zeta$. Under this assumption, the change of the probability distribution function due to diffusive motion may be governed by the multi-dimensional Smoluchowski's diffusion equation.

The bimolecular reaction process which is assumed to be irreversible can be described by introducing the pair absorbing probability as an additional term which expresses the probability that any pair of A and B particles disappears by forming the product, C. Thus, we have the equation:

$$\frac{\partial}{\partial t}P_{N,M}(X,\ Y;\ t\ |\ X_0,\ Y_0;\ 0)$$

$$= [D_A\nabla_X^2+D_B\nabla_Y^2-K(X,\ Y)]P_{N,M}(X,\ Y;\ t\ |\ X_0,\ Y_0;\ 0),$$

$$\nabla_X^2 = \sum_{i=1}^{N}\nabla_{xi}^2, \qquad \nabla_Y^2 = \sum_{j=1}^{M}\nabla_{yj}^2,$$

$$K(X,\ Y) = \sum_{i=1}^{N}\sum_{j=1}^{M}k(x_i,y_j). \tag{8.1}$$

This equation should be solved under the initial condition:

$$P_{N,M}(X, Y; 0 \mid X_0, Y_0; 0) = \delta(X-X_0)\delta(Y-Y_0). \qquad (8.2)$$

Equation 8.1 has the same form as the Bloch's equation for the density matrix of a many-body system in which $\hbar^2/2m$ and $1/kT$ correspond to the diffusion constant and the time, respectively. Therefore, we can adopt the various techniques developed in the field of many-body problems to find the approximate solutions of Eq. 8.1. In this paper, we obtained the solution of Eq. 8.1 in the form of a cluster expansion, which seems to be adequate for the present problem.

9. Cluster expansion

Now we shall derive a formula which expresses the reduced survival probability in a power series of densities of A and B particles. Here we are going to adopt the technique of cluster expansion, which was originally proposed by Kahn and Uhlenbeck (10) in relation to the problem of the condensation of a quantum gas, obeying the Boltzmanns' statistics.

Let us introduce a set of functions, $U_{k,l}(X, Y; t \mid X_0, Y_0; 0) = U_{k,l}$, which depends symmetrically on the coordinates of k A and l B particles in the volume V, respectively. They are defined by the following set of equations:

$$P_{1,0}(x; t \mid x_0; 0) = U_{1,0}(x; t \mid x_0; 0),$$
$$P_{0,1}(y; t \mid y_0; 0) = U_{0,1}(y; t \mid y_0; 0)$$
$$P_{1,1}(x, y; t \mid x_0, y_0; 0) = U_{1,1}(x, y; t \mid x_0, y_0; 0)$$
$$+ U_{1,0}(x, t \mid x_0; 0)U_{0,1}(y, t \mid y_0; 0),$$
$$P_{1,2}(x, y_1, y_2; t \mid x_0, y_{10}, y_{20}; 0) = U_{1,2}(x, y_1, y_2; t \mid x_0, y_{10}, y_{20}; 0)$$
$$+ U_{1,1}(x, y_1; t \mid x_0, y_{10}; 0)U_{0,1}(y_2; t \mid y_{20}; 0)$$
$$+ U_{1,1}(x, y_2; t \mid x_0, y_{20}; 0)U_{0,1}(y_1; t \mid y_{10}; 0)$$
$$+ U_{1,0}(x, t \mid x_0; t)U_{0,1}(y_1; t \mid y_{10}; 0)$$
$$\times U_{0,1}(y_2; t \mid y_{20}; 0), \qquad (9.1)$$

and so on. The general term is constructed by the following rule: the k A and l B particles are divided into a number of groups and the product of the U function for each group is formed. Then $P_{k,l}$ is constructed as the sum of those products for all possible ways of division of k A and l B particles. Next, we introduce the cluster integral, $b_{k,l}$, defined by

$$\int U_{k,l}(X, Y; t \mid X_0, Y_0; 0)\mathrm{d}X\mathrm{d}Y\mathrm{d}X_0\mathrm{d}Y_0 = Vk!\,l!\,b_{k,l}. \tag{9.2}$$

Using these definitions and integrating the expansion of Eq. 9.1 for $P_{N,M}$, we have the survival probability in the following form:

$$P(N, t) = \frac{N!\,M!}{V^{N+M}} \sum_{\lambda_{k,l}} \prod_{k,l} (Vb_{k,l})^{\lambda_{k,l}} / \lambda_{k,l}!\,, \tag{9.3}$$

where the summation is taken over all sets of values of $\lambda_{k,l}$, which satisfy the conditions:

$$\sum_{k,l} k\lambda_{k,l} = N\,, \qquad \sum_{k,l} l\lambda_{k,l} = M\,. \tag{9.4}$$

Equation 9.3 is too complicated to use in performing the calculation. However, if we consider the case where $N+M\to\infty$ and $v=V/(N+M)$ has a finite value, Eq. 9.3 is reduced to a simpler form by using the theorem of Biermann and Lemaire (3, 14). Let us consider the double power series:

$$F(a, b) = \sum_{N,M} Q_{N,M} a^N b^M\,. \tag{9.5}$$

The theorem of Biermann and Lemaire states that, if positive power series $F(a, b)$ is convergent, the following equation holds

$$\limsup_{N+M\to\infty} {}^{N+M}\sqrt{Q_{N,M} a_0{}^N b_0{}^M} = 1\,, \tag{9.6}$$

where a_0 and b_0 are associated radii of the convergence of $F(a, b)$. Now, as the coefficients $Q_{N,M}$, we shall take the right-hand side of Eq. 9.3:

$$Q_{N,M} = \sum_{\lambda_{k,l}} \prod_{k,l} \{(N+M)vb_{k,l}\}^{\lambda_{k,l}} / \lambda_{k,l}!\,, \tag{9.7}$$

Then, by substituting the associated radii which can be calculated from Eq. 9.5, into Eq. 9.6, we can obtain $Q_{N,M}$ or the reduced survival probability (9.3) for the limiting case $N+M\to\infty$, $V\to\infty$.

Noting that $Q_{N,M}$ is the coefficient of $a^N b^M$ of the function $\exp\{\sum_{k,l}(N+M)vb_{k,l}\,a^k\beta^l\}$, we can write Eq. (9.5) by the Cauchy's integral:

$$F(a, b) = \left(\frac{1}{2\pi_i}\right)^2 \sum_{N,M} \oint\oint \exp\left\{\sum_{k,l}(N+M)vb_{k,l}\alpha^k\beta^l\right\}$$

$$\times \left(\frac{a}{\alpha}\right)^N \left(\frac{b}{\beta}\right)^M \frac{1}{\alpha\beta} \mathrm{d}\alpha\mathrm{d}\beta , \tag{9.8}$$

where the integral has to be taken around the origin of the complex α, β plane, excluding the singularities of $\sum_{k,l} vb_{k,l}\,\alpha^k\beta^l$. The integral can be carried out by applying the theorem of residues as

$$F(a,b) = 1/\{1 - \sum_{k,l} vkb_{k,l}\alpha^k\beta^l - \sum_{k,l} vlb_{k,l}\alpha^k\beta^l\} , \tag{9.9}$$

where α and β appearing in the sum of Eq. 9.9 have values determined by the equations:

$$\alpha = a \exp \{\sum_{k,l} vb_{k,l}\alpha^k\beta^l\}$$

$$\beta = b \exp \{\sum_{k,l} vb_{k,l}\alpha^k\beta^l\} . \tag{9.10}$$

Then we shall obtain the line of the nearest singularity of $F(a, b)$ which determines the associated radii of convergence, a_0 and b_0.

One can see from Eqs. 9.9 and 9.10 that the possible singularities of $F(a, b)$ are determined by the values of α_0 and β_0 which correspond to the singularity of $\sum_{k,l} vb_{k,l}\,\alpha^k\beta^l$ and also the zero point of denominator of Eq. 9.9:

$$1 - \sum_{k,l} vkb_{k,l}\alpha^k\beta^l - \sum_{k,l} vlb_{k,l}\alpha^k\beta^l = 0 . \tag{9.11}$$

If v is assumed to be sufficiently large, the nearer singularity may be that of the zero point of $1 - \sum_{k,l} vkb_{k,l}\alpha^k\beta^l - \sum_{k,l} vlb_{k,l}\alpha^k\beta^l$.

Returning to the theorem of Biermann and Lemaire, we can translate it into our nomenclature as follows. If we assume that $\lim_{N,M\to\infty} \{1/(N+M)$ $\log Q_{N,M} = \xi(x)$ can be defined as a continuous function of the ratio $x = N/(N+M)$, then we have

$$\mathrm{Max}_{x} \{\xi(x) + x \log a_0 + (1-x) \log b_0\} = 0 . \tag{9.12}$$

This is rewritten at this maximum point as

$$\xi(x) + x \log a_0 + (1-x) \log b_0 = 0$$

$$\frac{\mathrm{d}}{\mathrm{d}x} \xi(x) + \log a_0 - \log b_0 = 0 . \tag{9.13}$$

Substituting Eq. 9.10 into the above equations and taking into account Eq. 9.11, we can obtain the relations (26):

$$N/V = \sum_{k,l} k b_{k,l} \alpha_0{}^k \beta_0{}^l , \qquad M/V = \sum_{k,l} l b_{k,l} \alpha_0{}^k \beta^l . \tag{9.14}$$

Now $\log P(N, t)$ is obtained by introducing Eq. 9.10 into Eq. 9.13:

$$\log P(N,t) = \log N! \, M!/V + V \sum_{k,l} b_{k,l} \alpha_0{}^k \beta_0{}^l - N \log \alpha_0 - M \log \beta_0 , \tag{9.15}$$

where the parameters α_0 and β_0 are determined by Eq. 9.14. In order to find an explicit expression of $P(N, t)$ as a function of the concentrations of the A and B particles, N/V and M/V, we may carry out a successive approximation. Thus, finally, the survival probability in the power series of concentration up to the fourth order is calculated as

$$
\begin{aligned}
P(N, t) = \exp\Big[&\frac{NM}{V} b_{1,1} - \frac{1}{V^2}\Big\{ \frac{1}{2} NM(N+M)b^2{}_{1,1} - N^2 M b_{2,1} - NM^2 b_{1,2} \Big\} \\
&+ \frac{1}{V^3}\Big\{ \frac{1}{3} NM(N^2+3NM+M^2)b^3{}_{1,1} - N^3 M b_{1,1} b_{2,1} - 2N^2 M^2 b_{1,1} \\
&\times (b_{1,2}+b_{2,1}) - NM^3 b_{1,1} b_{1,2} + N^3 M b_{3,1} + N^2 M^2 b_{2,2} + NM^3 b_{1,3} \Big\} \\
&+ \cdots \Big].
\end{aligned}
\tag{9.16}
$$

10. *Binary collision expansion*

The survival probability function $P_{k,l}(X, Y; t|X_0, Y_0; 0)$ is shown to fulfill Eq. 8.1, which has the same form as the Bloch's equation, so we shall first obtain the solution by adopting the binary collision expansion method which was introduced by Yang and Lee (*13*), and Siegert and Teramoto (*28*) for the calculation of the density matrix of a many-body system. Then by inserting the solutions thus obtained into Eq. 9.1, we will have the function $U_{k,l}$ in terms of binary collision expansion.

Simply rewriting the right-hand side of Eq. 8.1 with an operator $\mathscr{D}$, we have

$$\frac{\partial}{\partial t} P = \mathscr{D} P , \tag{10.1}$$

where

$$\mathscr{D} = \mathscr{D}^0 - K , \qquad \mathscr{D}_0 = D_A \sum_{i=1}^{N} \nabla_{xi}{}^2 + D_B \sum_{j=1}^{M} \nabla_{yj}{}^2 ,$$

$$K = \sum_{\alpha} k_\alpha = \sum k(r_\alpha), \tag{10.2}$$

and P is the abbreviated notation of $P_{N,M}(X, Y; t|X_0, Y_0; t_0)$.

In the last equation of Eq. 10.2, the pairs of A and B particles are numbered by small Greek subscripts. The sum extends over all the AB pairs and r_α is the relative distance of the pair α. The function P is understood to be the principal solution which satisfies the boundary condition of infinite volume and the initial condition (8.2).

Now, if we consider the equation:

$$\frac{\partial}{\partial t}P_\alpha = (\mathscr{D}_0 - k_\alpha)P_\alpha = \mathscr{D}_\alpha P_\alpha, \tag{10.3}$$

the principal solution P_α is related to P by the integral equation:

$$\begin{aligned}
P(X, Y; t \mid X_0, Y_0; t_0) = &\, P_\alpha(X, Y; t \mid X_0, Y_0; t_0) \\
&- \int_{t_0}^{t} dt' \int dX' dY' P_\alpha(X, Y; t \mid X', Y'; t') \\
&\times \sum_{\beta, \beta \neq \alpha} k(x_\beta', y_\beta')P(X', Y'; t' \mid X_0, Y_0; t_0). \quad (10.4)
\end{aligned}$$

We shall introduce an abbreviated notation $\{F, k_\alpha G\}$ defined by

$$\begin{aligned}
\{F, k_\alpha G\} = &\int_{t_0}^{t} dt' \int dX' dY' F(X, Y; t \mid X', Y'; t')k(x_\alpha', y_\alpha') \\
&\times G(X', Y'; t' \mid X_0, Y_0; t_0).
\end{aligned}$$

In this notation, Eq. 10.4 is written as

$$P = P_\alpha - \{P_\alpha, \sum_{\beta, \beta \neq \alpha} k_\beta P\}. \tag{10.5}$$

Furthermore, we have the well-known equation:

$$P = P_0 - \{P_0, \sum_{\alpha} k_\alpha P\}, \tag{10.6}$$

where P_0 is the principal solution of the multi-dimensional diffusion equation for the nonreacting particle system ($\sum k_\alpha = 0$), so that it is given in the form of Gaussian distribution with the variance Dt.

Starting from Eq. 10.5 and 10.6, we can find a successive approximation which results in the binary collision expansion:

$$P = P_0 - \sum_{\alpha} \{P_0, k_\alpha P_\alpha\} + \sum_{p \neq \alpha} \{P_0, k_\alpha P_\alpha, k_\beta P_\beta\}$$

$$- \sum_{\substack{\beta \neq \alpha \\ \gamma \neq \beta}} \{P_0, k_\alpha P_\alpha, k_\beta P_\beta, k_\gamma P_\gamma\} + \cdots , \tag{10.7}$$

where the symbols $\sum\limits_{\alpha \neq \beta}$, $\sum\limits_{\beta \neq \alpha,\, \gamma \neq \beta}$, *etc.* mean the summation of all the indices, excluding the values of $\beta = \alpha$, $\gamma = \beta$. If the pair distribution function is defined by

$$p(x_\alpha, y_\alpha; t \mid x_{\alpha 0}, y_{\alpha 0}; 0) = \int P_\alpha(X, Y; t \mid X_0, Y_0; 0) d\underline{X} d\underline{Y},$$

where $\underline{X}$ and $\underline{Y}$ denote all the coordinates X and Y, except for those of the pair, x_α and y_α, it satisfies the equation:

$$\frac{\partial}{\partial t} p(x_\alpha, y_\alpha; t \mid x_{\alpha 0}, y_{\alpha 0}; 0) = \{D_A \nabla_{x\alpha}^2 + D_B \nabla_{y\alpha}^2 - k_\alpha(x_\alpha, y_\alpha)\}$$

$$\times p(x_\alpha, y_\alpha; t \mid x_{\alpha 0}, y_{\alpha 0}; 0). \tag{10.8}$$

It is easily seen from Eq. 10.3 that P_α and $k_\alpha P_\alpha$ can be written by means of the pair distribution function as

$$P_\alpha = p(x_\alpha, y_\alpha; t \mid x_{\alpha 0}, y_{\alpha 0}; 0) P_0(\underline{X}, \underline{Y}; t \mid \underline{X}_0, \underline{Y}_0; 0), \tag{10.9}$$

$$k_\alpha P_\alpha = k(x_\alpha, y_\alpha) p(x_\alpha, y_\alpha; t \mid x_{\alpha 0}, y_{\alpha 0}; 0) P_0(\underline{X}, \underline{Y}; t \mid \underline{X}_0, \underline{Y}_0; 0). \tag{10.10}$$

Thus we see that the basic functions of the binary collision expansion are p_α and $k_\alpha p_\alpha$. Hereafter, we shall designate the latter as the binary kernel. If we introduce new variables, $\mathbf{r}$ and $\mathbf{R}$, which are defined by

$$\mathbf{r} = x_\alpha - y_\alpha ,$$

$$\mathbf{R} = (D_B x_\alpha + D_A y_\alpha)/(D_A + D_B), \tag{10.11}$$

and assume that k_α depends only on the relative distance r, the pair distribution function can be written in the form:

$$p_\alpha = g(\mathbf{R}, \mathbf{R}_0; t) f(\mathbf{r}, \mathbf{r}_0; t), \tag{10.12}$$

where g and f are the solutions to the equations:

$$\frac{\partial}{\partial t} g = D_R \nabla^2 g$$

$$\frac{\partial}{\partial t} f = (D_r \nabla^2 - k_\alpha) f, \quad D_R = D_A D_B/(D_A + D_B), \quad D_r = D_A + D_B. \tag{10.13}$$

Now we can obtain the U function in terms of the binary kernel. In Eq. 9.1, $U_{k,l}$ is expressed uniquely as

$$
\begin{aligned}
U_{1,1}(x, y\,;\, t\,|\,x_0, y_0\,;\, 0) = {} & P_{1,1}(x, y\,;\, t\,|\,x_0, y_0\,;\, 0) \\
& - P_{1,0}(x\,;\, t\,|\,x_0\,;\, 0)P_{0,1}(y\,;\, t\,|\,y_0\,;\, 0)\,, \\
U_{1,2}(x, y_1, y_2\,;\, t\,|\,x_0, y_{10}, y_{20}\,;\, 0) = {} & P_{1,2}(x, y_1, y_2\,;\, t\,|\,x_0, y_{10}, y_{20}\,;\, 0) \\
& - P_{1,1}(x, y_1\,;\, t\,|\,x_0, y_{10}\,;\, 0)P_{0,1}(y_2\,;\, t\,|\,y_{20}\,;\, 0) \\
& - P_{1,1}(x, y_2\,;\, t\,|\,x_0, y_{20}\,;\, 0)P_{0,1}(y_1\,;\, t\,|\,y_{10}\,;\, 0) \\
& + P_{1,0}(x\,;\, t\,|\,x_0\,;\, 0)P_{0,1}(y_1\,;\, t\,|\,y_{10}\,;\, 0) \\
& \times P_{0,1}(y_2\,;\, t\,|\,y_{20}\,;\, 0)\,, \ldots\ldots\,. \qquad (10.14)
\end{aligned}
$$

Inserting the binary collision expansion of P into the above equations, we obtain $U_{k,l}$ as the sum of all the possible sequences of the binary kernels in which all particles are connected at least once through "lines." If we represent the binary kernels by the "lines" linking the corresponding pairs of A and B particles, we have

$$
\begin{aligned}
U_{1,1} = {} & -\{P_0, k_\alpha P_\alpha\}, \\
U_{2,1} = {} & \sum_{\beta\neq\alpha} \{P_0, k_\alpha P_\alpha, k_\beta P_\beta\} - \sum_{\beta\neq\alpha} \{P_0, k_\alpha P_\alpha, k_\beta P_\beta, k_\alpha P_\alpha\} + \cdots, \\
U_{2,2} = {} & \sum_{\substack{\beta\neq\alpha \\ \gamma\neq\beta,\alpha}} \{P_0, k_\alpha P_\alpha, k_\beta P_\beta, k_\gamma P_\gamma\} + \sum \{P_0, k_\alpha P_\alpha, k_\beta P_\beta, k_\gamma P_\gamma, k_\delta P_\delta\} - \cdots,
\end{aligned}
$$

$$
\ldots\ldots\,, \qquad (10.15)
$$

where the summation of the second term of $U_{2,2}$ is taken from those sequences which fulfill either of the following sets of conditions:

$$
\beta\neq\alpha, \gamma\neq\beta,\alpha,\ \delta\neq\gamma \quad\text{or}\quad \beta\neq\alpha, \gamma=\alpha, \delta\neq\alpha,\beta\,.
$$

In the remaining part of this section, we are going to introduce the momentum representation in order to simplify the calculation of the cluster integral in Eq. 9.2. Here we assume $V\rightarrow\infty$ as before. So Eq. 9.2 can be written as

$$
b_{k,l} = \frac{1}{k!l!} \int U_{k,l}(X, Y\,;\, t\,|\,0, x_{20}, x_{30}, \ldots, x_{k0}, Y_0\,;\, 0)\mathrm{d}X\mathrm{d}Y\mathrm{d}\underline{X}_0\mathrm{d}Y_0\,. \quad (10.16)
$$

Let us introduce the momentum representation $F(K_X, K_Y\,;\, t\,|\,K_{X0}, K_{Y0}\,;\, 0)$ of the function $F(X, Y\,;\, t\,|\,X_0, Y_0\,;\, 0)$, which is defined by

$$
F(X, Y\,;\, t\,|\,X_0, Y_0\,;\, 0) = \left(\frac{1}{8\pi}\right)^{k+l} \int \mathrm{d}K_X\mathrm{d}K_Y\mathrm{d}K_{X0}\mathrm{d}K_{Y0}
$$

$$\times \exp \{i \sum k_{x_i} x_i + i \sum k_{y_j} y_j - i \sum k_{x_{i0}} x_{i0}$$
$$- i \sum k_{y_{j0}} y_{j0}\} F(K_X, K_Y; t \mid K_{X0}, K_{Y0}; 0), \quad (10.17)$$

where $K_X = (k_{x1}, k_{x2}, \ldots, k_{xk})$ and $K_Y = (k_{y1}, k_{y2}, \ldots, k_{yl})$ are sets of momenta for k A and l B particles, respectively. As a result of the conservation of momentum for the operator $\mathscr{D}^0 - \sum_{i,j} k(|x_i - y_j|)$, the momentum representation of $U_{k,i}$ can be written in the form of the product of a delta function and a certain function u, which is

$$U_{k,l}(K_X, K_Y; t \mid K_{X0}, K_{Y0}; 0) = \delta(\sum k_{x_i} + \sum k_{y_j} - \sum k_{x_i} - \sum k_{y_j})$$
$$\times u_{k,l}(K_X, K_Y; t \mid K_{X0}, K_{Y0}; 0). \quad (10.18)$$

Taking into account this relationship and Eq. (10.17), we can carry out the integration of Eq. 10.16. Then b_{kl} can be expressed, using the function u at $K_X = K_Y = K_{X0} = K_{Y0} = 0$, as

$$b_{k,l} = \frac{(8\pi^3)^{k+l-1}}{k! l!} u_{k,l}(0, 0; t \mid 0, 0; 0). \quad (10.19)$$

The rule of expansion in deriving Eqs. 10.7 and 10.15 remains valid also for the momentum representation, so we have only to exchange each $k_\alpha P_\alpha$ and P_α of Eqs. 10.7 and 10.15 with their momentum representations.

In the following section, we shall give the explicit operator forms for $U_{1,1}$ and $k_\alpha p_\alpha$ and their momentum representation. From Eq. 10.14 the operator form of $U_{1,1}$ is

$$U_{1,1}(t) = \exp (Dt_A \nabla_x^2 + D_B t \nabla_y^2 - k(x, y)t)$$
$$- \exp (D_A t \nabla_x^2) \exp (D_B t \nabla_y^2). \quad (10.20)$$

Taking the time derivative of the above equation and using Eq. 10.8, we can obtain the operator form of the binary kernel:

$$k_\alpha p_\alpha = -\frac{\partial}{\partial t} U_{1,1}(t) + (D_A \nabla_x^2 + D_B \nabla_y^2) U_{1,1}(t). \quad (10.21)$$

According to these relations, their momentum representation is as follows:

$$U_{1,1}(k_x, k_y; t \mid k_{x0}, k_{y0}; 0) = p_\alpha(k_x, k_y; t \mid k_{x0}, k_{y0}; 0)$$
$$- P_0(k_x, k_y; t \mid k_{x0}; k_{y0}; 0),$$

$$k_\alpha p_\alpha(k_x, k_y\,;\,t\,|\,k_{x0}, k_{y0}\,;\,0) = -\frac{\partial}{\partial t} U_{1,1}(k_x, k_y\,;\,t\,|\,k_{x0}, k_{y0}\,;\,0)$$

$$+(D_A k_x{}^2 + D_B k_y{}^2)$$

$$\times U_{1,1}(k_x, k_y\,;\,t\,|\,k_{x0}, k_{y0}\,;\,0)\,, \qquad (10.22)$$

where

$$P_0(k_x, k_y\,;\,t\,|\,k_{x0}, k_{y0}\,;\,0)$$

$$= \delta(k_x - k_{x0})\delta(k_y - k_{y0}) \exp\left(-D_A t k_x{}^2 - D_B t k_y{}^2\right).$$

11.   *Binary kernels in momentum representation*

Now we shall solve Eq. 10.8 for the two kinds of diffusion-controlled reactions, corresponding to the so-called 1) Smoluchowski and 2) radiation boundary conditions (hereafter, we will refer to these two cases simply as s.b.c. and r.b.c., respectively) and obtain the binary kernels in momentum representation for each case.

Firstly, we shall consider the case of s.b.c., where the reaction between the A and B particles occurs immediately after the relative distance between the A and B particles comes within the range σ. This corresponds to the assumption that the walls on the sphere of radius σ are completely absorbent. Therefore, $k(x, y)$ for s.b.c. can be written as

$$k(x, y) = \infty \qquad \text{for} \qquad |x-y| \leq \sigma\,,$$

$$k(x, y) = 0 \qquad \text{for} \qquad |x-y| > \sigma\,. \qquad (11.1)$$

By putting these relations into Eq. 10.8, which has the same form as the Shrödinger's equation for two interacting particles of inert gas with collision radius σ, it can be shown that the explicit solutions g and f become

$$g(\mathbf{R}, \mathbf{R}_0\,;\,t) = \frac{1}{(4\pi D_R t)^{3/2}} \exp \frac{-(\mathbf{R}-\mathbf{R}_0)^2}{4D_R t}\,,$$

$$f(\mathbf{r}, \mathbf{r}_0\,;\,t) = \frac{2}{\pi r_0 r} \sum_l \sum_m Y_{lm}(\theta, \phi) Y_{lm}{}^*(\theta_0, \phi_0)$$

$$\times \int_0^\infty dk Z_{k,l}(r) Z_{k,l}{}^*(r_0) \exp\left(-k^2 D_r t\right)\,, \qquad (11.2)$$

where we introduce the spherical coordinates r, θ, ϕ, instead of the vector $\mathbf{r}$ and

$$Z_{k,l}(r) = \sqrt{\frac{\pi k r}{2}} \frac{J_{-l-1/2}(k\sigma)J_{l+1/2}(kr) - J_{l+1/2}(k\sigma)J_{-l-1/2}(kr)}{\{J^2_{l+1/2}(k\sigma) + J^2_{-l-1/2}(k\sigma)\}^{1/2}}, \tag{11.3}$$

and Y_{lm}, $J_{l+1/2}$ are the normalized spherical harmonics and the Bessel's function of order $l+1/2$, respectively. Then the functions $U_{k,l}$ can be explicitly given by inserting these solutions into Eq. 10.14.

Next, in the calculation of the cluster integrals $b_{k,l}$ that is, Eq. 9.2, if we integrate first over angular variables, considering the properties of spherical harmonics:

$$\int_0^{2\pi} \int_{-1}^{1} Y_{lm}(\theta, \phi)\,\mathrm{d}(\cos\theta)\,\mathrm{d}\phi = 2\pi \qquad \text{for} \qquad l = m = 0,$$

$$= 0 \qquad \text{otherwise},$$

we can see that we have only to consider the contribution of s state ($l=m=0$) for the evaluation of $b_{k,l}$. So we shall calculate the binary kernel of s state, following the procedure presented above. From Eq. 11.3, we have

$$Z_{0,0}(r) = \sin(kr - k\sigma) \qquad \text{for} \quad r > \sigma,$$

$$= 0 \qquad \text{for} \quad r \leq \sigma. \tag{11.4}$$

Substituting these relations into Eq. 11.2 and using Eq. 10.15, we have

$$U_{1,1} = \frac{1}{4\pi r r_0 (4\pi D_r t)^{1/2}(4\pi D_R t)^{3/2}} \exp\frac{-(\mathbf{R}-\mathbf{R}_0)^2}{4D_R t}\left\{\exp\frac{-(r+r_0)^2}{4D_r t}\right.$$

$$\left. -\exp\frac{-(r+r_0-2\sigma)^2}{4D_r t}\right\} \qquad \text{for} \quad r > \sigma,\ r_0 > \sigma,$$

$$= \frac{1}{4\pi r r_0 (4\pi D_r t)^{1/2}(4\pi D_R t)^{3/2}} \exp\frac{-(\mathbf{R}-\mathbf{R}_0)^2}{4D_R t}\left\{\exp\frac{-(r+r_0)^2}{4D_r t}\right.$$

$$\left. -\exp\frac{-(r-r_0)^2}{4D_r t}\right\} \qquad \text{otherwise}. \tag{11.5}$$

Then, from the definition of the momentum representation, Eq. 10.17 and Eq. 10.22, we can obtain the binary kernel in k representation in the form:

$$k_\alpha p_\alpha = \frac{-1}{2\pi^2 k_0 k(k_0^2 - k^2)} \delta(\mathbf{K} - \mathbf{K}_0)\exp(-D_R t\mathbf{K}_0^2)(D_A k_x^2 + D_B k_y^2$$

$$-D_R\mathbf{K}^2-D_r k^2)\Big[k_0 \cos k_0\sigma \sin k\sigma \exp\left(-D_r t k_0^2\right)$$

$$-k \sin k_0\sigma \cos k\sigma \exp\left(-D_r t k^2\right)-\frac{2}{\sqrt{\pi}}\sin k_0\sigma$$

$$\times \sin k_0\sigma \{k_0 M(\sqrt{D_r t}\,k_0)\exp\left(-D_r t k_0^2\right)-kM(\sqrt{D_r t}\,k)$$

$$\times \exp\left(-D_r t k^2\right)\}\Big]+\frac{D_r}{2\pi^2 k_0 k}\delta(\mathbf{K}-\mathbf{K}_0)\exp\left(-D_R t\mathbf{K}_0^2\right)$$

$$\times\left\{k_0 \cos k_0\sigma \sin k\sigma \exp\left(-D_r t k_0^2\right)-\frac{2}{\sqrt{\pi}}k_0 \sin k_0\sigma \sin k\sigma\right.$$

$$\left.\times M(\sqrt{D_r t}\,k_0)\exp\left(-D_r t k_0^2\right)+\frac{1}{\sqrt{\pi D_r t}}\sin k_0\sigma \sin k\sigma\right\}$$

$$+\frac{1}{2\pi^2 k_0 k(k_0^2-k^2)}\delta(t)\delta(\mathbf{K}-\mathbf{K}_0)$$

$$\times \{k_0 \cos k_0\sigma \sin k\sigma -k \sin k_0\sigma \cos k\sigma\}\,, \tag{11.6}$$

where

$$k_x+k_y=\mathbf{K}\,,\quad |k_x-k_y|=2k \quad\text{and}\quad M(\alpha)=\int_0^\alpha \exp(a^2)\mathrm{d}a\,.$$

If we put $D_\mathrm{A}=D_\mathrm{B}$ here, we can simplify the form as

$$k_\alpha p_\alpha=\frac{D_r}{2\pi^2 k_0 k}\delta(\mathbf{K}-\mathbf{K}_0)\exp\left(-D_R t\mathbf{K}_0^2-D_r t k_0^2\right)\sin k\sigma\left\{k_0 \cos k_0\sigma\right.$$

$$\left.-\frac{2}{\sqrt{\pi}}k_0 \sin k_0\sigma M(\sqrt{D_r t}\,k_0)+\frac{1}{\sqrt{\pi D_r t}}\sin k_0\sigma \exp\left(D_r t k_0^2\right)\right\}$$

$$+\frac{1}{2\pi^2 k_0 k(k_0^2-k^2)}\delta(t)\delta(\mathbf{K}-\mathbf{K}_0)$$

$$\times(k_0 \cos k_0\sigma \sin k\sigma -k \sin k_0\sigma \cos k\sigma)\,. \tag{11.7}$$

In order to obtain $U_{k,l}$ in k representation, we insert Eq. 11.7 into Eq. 10.15. In the integration with k, the effective values of k will be limited within the range $0<k<\dfrac{1}{\sqrt{D_r t}}$, due to the exponential factor in Eq. 11.7. Therefore, if $\sigma/\sqrt{D_r t}<1$, the main contribution of the first, second and third terms in the bracket of the first term of Eq. 11.7 is estimated to be in the order of $1/\sqrt{D_r t}$, $\sigma/D_r t$ and $\sigma/D_r t$, respectively.

In the next place, we shall obtain the binary kernel for the case of r.b.c.   The calculation is performed in a manner similar to the one used in the case of s.b.c.   The equation for the pair distribution function of r.b.c. is written

$$\frac{\partial}{\partial t}f = D_r{}^2\nabla f, \qquad D_r\frac{\partial}{\partial r}f = \beta f \qquad \text{as} \qquad r \to \sigma, \tag{11.8}$$

where the latter equation, which is introduced in place of the pair absorbing potential $k(r)$, indicates a radiation boundary condition where the rate of reaction on the sphere of radius σ is proportional to the concentration of counter particles on the sphere.   This equation can be reduced to the s.b.c. equation, if we make $\beta \to \infty$.   The principal solution of Eq. 11.8 is calculated as

$$
\begin{aligned}
p_\alpha(\mathbf{r}, \mathbf{R}\,;\, t \mid \mathbf{r}_0, \mathbf{R}_0\,;\, 0) = {}& \frac{1}{(4\pi D_R t)^{3/2}} \exp \frac{-(\mathbf{R}-\mathbf{R}_0)^2}{4D_R t} \left[\frac{1}{8\pi r_0 r \sqrt{\pi D_r t}} \right. \\
& \times \left\{ \exp -\frac{(r-r_0)^2}{4D_r t} + \exp -\frac{(r+r_0-2\sigma)^2}{4D_r t} \right\} \\
& -\frac{h}{4\pi r r_0}\, \exp\left(D_r t h^2 + h(r+r_0-2\sigma)\right) \\
& \left. \times \operatorname{erfc}\left\{ \frac{r+r_0-2\sigma}{\sqrt{4D_r t}} + h\sqrt{D_r t} \right\} \right],
\end{aligned}
\tag{11.9}
$$

where

$$\operatorname{erfc}\alpha = \frac{2}{\sqrt{\pi}} \int_\alpha^\infty \exp(-a^2)\mathrm{d}a, \qquad h = \frac{1}{\sigma} + \frac{\beta}{D_r}.$$

For the same reason as mentioned in the case of s.b.c., we considered only s state.

Now we can obtain the binary kernel in momentum representation by the same procedure we used in the case of s.b.c.   The results are summarized as

$$
\begin{aligned}
k_\alpha p_\alpha = {}& \frac{-D_r}{2\pi k_0 k}\delta(\mathbf{K}-\mathbf{K}_0) \exp(-D_R t \mathbf{K}^2)\left[\frac{k_0}{k_0{}^2+h^2} \{(kk_0 \sin k_0\sigma \cos k\sigma \right. \\
& -h^2 \sin k\sigma \cos k_0\sigma) + h(k \cos k\sigma \cos k_0\sigma - k_0 \sin k\sigma \sin k_0\sigma)\} \\
& \times \exp(-D_r t k_0{}^2) + \frac{1}{k_0{}^2+h^2}\left\{ h \exp(D_r t h^2) \operatorname{erfc}(h\sqrt{D_r t}) \right. \\
& \left. -\frac{2k_0}{\sqrt{\pi}} \exp(-D_r t k_0{}^2) M(k_0\sqrt{D_r t}) \right\} \{h(k \cos k\sigma \sin k_0\sigma
\end{aligned}
$$

$$+k_0 \sin k\sigma \cos k_0\sigma)-k_0 k \cos k_0\sigma \cos k\sigma-h^2 \sin k\sigma \sin k_0\sigma\} \Bigg]$$

$$+\frac{1}{2\pi^2 k_0 k(k^2-k_0^2)}\delta(t)\delta(\mathbf{K}-\mathbf{K}_0)$$

$$\times \{k_0 \cos k_0\sigma \sin k\sigma-k \cos k\sigma \sin k_0\sigma\}\,, \tag{11.10}$$

where we have assumed $D_A{=}D_B$ to simplify the calculation. As expected, this equation can be reduced to the binary kernel of s.b.c., Eq. 11.7, if we make $\beta{\to}\infty$.

12. *Approximate results*

Now we shall calculate $u_{k,l}(0, 0;\ t|0, 0;\ 0)$ from Eq. 10.19 by inserting the binary kernels obtained in the last section into each term of Eq. 10.15 in momentum representation and then integrating with the momentum.   Furthermore, using these results in Eq. 9.16, we can obtain the survival probabilities.   Here we are going to consider a special case, $D_A{=}D_B$, and calculate $u_{k,l}$ for the cluster of two, three and four particles in both the cases of s.b.c. and r.b.c.

Firstly, we consider the case of s.b.c.   The function $u_{1,1}$ for the cluster containing one A and one B particle can be obtained exactly. It is

$$u_{1,1}(0,0;\ t\,|\,0,0;\ 0)= -\frac{D_r\sigma t}{2\pi^2}\left(1+\frac{2\sigma}{\sqrt{\pi D_r t}}\right). \tag{12.1}$$

Inserting this equation into Eq. 10.19, we obtain the cluster integral for two particles.   It is

$$b_{1,1}= -4\pi D_r\sigma t\left(1+\frac{2\sigma}{\sqrt{\pi D_r t}}\right).$$

By comparing this result with Eq. 5.2, it is seen that $db_{1,1}/dt$ corresponds to the reaction rate of the two particle system, which has already been derived by many authors.

In the next place, let us consider the cluster of three particles and the function $u_{1,2}{=}u_{2,1}$.   In the binary collision expansion of $u_{1,2}$, Eq. 10.15, the first term which contains two binary kernels, can be obtained exactly; but the other terms which contain more than two binary kernels are very difficult to calculate.   Here, taking into account the order estimation of each term of the binary kernel of Eq. 11.7, which was discussed in the last section, we can obtain $u_{1,2}=u_{2,1}$ in a power series of $\sigma/\sqrt{D_r t}$, under the condition $\sigma/\sqrt{D_r t}<1$, as

$$u_{1,2}(0,0;t\,|\,0,0;0)=\left(\frac{D_r\sigma t}{2\pi^2}\right)^2\left\{1+4\frac{\sigma}{\sqrt{\pi D_r t}}+(2-2\alpha_1+2\alpha_2\right.$$

$$\left.+\frac{\sqrt{3}}{\pi}\log\frac{\sigma^2}{D_r t}\right)\frac{\sigma^2}{D_r t}+\cdots\right\},\tag{12.2}$$

where

$$\alpha_1=\frac{-1}{\pi}\left(\frac{\sqrt{3}}{2}\gamma-\frac{5}{2\sqrt{3}}-\frac{\pi}{4}+\frac{7\sqrt{3}}{4}\log 3-\frac{4}{\sqrt{3}}\log 2\right),$$

$$\alpha_2=\left(\frac{2}{\pi}\right)^2\int_0^\infty dx\int_0^\infty dy\int_0^1 d(\cos\theta)\frac{\sin y\cos x\sin\left(\frac{x}{2}\right)\cos\left(\frac{y}{2}\right)}{(x^2+xy\cos\theta+y^2)}$$

$$\times\frac{\sin(x^2+xy\cos\theta+y^2/4)^{1/2}\cos(y^2+xy\cos\theta+y^2/4)^{1/2}}{\times(x^2+xy\cos\theta+y^2/4)^{1/2}xy},$$

and γ is the Euler's constant, *i.e.*, $\gamma=0.57721\ldots$ Similarly we can obtain $u_{2,2}$ and $u_{1,3}=u_{3,1}$, whose clusters contain four particles, in the forms:

$$u_{2,2}(0,0;t\,|\,0,0;0)=\left(\frac{D_r\sigma t}{2\pi^2}\right)^3\left\{-4-\left(\frac{248}{15}+\frac{64}{15}\sqrt{2}\right)\frac{\sigma}{\sqrt{\pi D_r t}}+\cdots\right\},$$

$$u_{1,3}(0,0;t\,|\,0,0;0)=\left(\frac{D_r\sigma t}{2\pi^2}\right)^3\left\{-1-6\frac{\sigma}{\sqrt{\pi D_r t}}+\cdots\right\}.\tag{12.3}$$

In general, it can be shown that $u_{k,l}$ for the cluster of k A and l B particles is also expanded in the power series of $\sigma/\sqrt{D_r t}$ whose lowest order term is given by

$$\frac{\left(\dfrac{-D_r\sigma t}{2\pi^2}\right)^{k+l-1}}{(k+l-1)!}\times\left\{\begin{array}{l}\text{The number of possible arrangements in the con-}\\ \text{nected clusters containing the smallest number of}\\ \text{binary kernels, or the number of terms in the sum-}\\ \text{mation of the first term of }U_{k,l},\text{ (10.15).}\end{array}\right\}$$

Substituting these results into Eqs. 10.19 and 9.16, we finally obtain the survival probability in the form of a density expansion up to the fourth order:

$$P(N,t)=\exp\left\{-\left(1+\frac{2\sigma}{\sqrt{\pi D_r t}}\right)(4\pi D_r\sigma t)\frac{NM}{V}\right.$$

$$\left.-\left(\frac{2}{\pi}-1+\alpha_1-\alpha_2-\frac{\sqrt{3}}{2\pi}\log\frac{\sigma^2}{D_r t}\right)\left(\frac{\sigma^2}{D_r t}\right)(4\pi D_r\sigma t)^2\frac{(N+M)NM}{V^2}\right.$$

$$+\left(\frac{28}{15}-\frac{16\sqrt{2}}{15}\right)\frac{\sigma}{\sqrt{\pi D_r t}}(4\pi D_r\sigma t)^3\frac{N^2 M^2}{V^3}+\cdots\right\}. \tag{12.4}$$

This equation holds true under the condition $\sigma^2/D_r<t<V/(D_r\sigma(N+M))$. The coefficient of the second term in the exponent, that is, the contribution from the three particle clusters, begins with a term whose order is higher than the expected order:

$$\frac{\sigma}{\sqrt{\pi D_r t}}(4\pi D_r\sigma t)^2\frac{NM(N+M)}{V^2},$$

because it comes from the clusters of two A and one B or one A and two B particles. These clusters have reducible forms because of the lack of interaction between two particles of the same species (15).

Now we are going to discuss some important predictions that we can make, based on this result. In order to simplify the discussion, we shall consider the special case that $N=M$ and define the average survival probability for one particle at a constant density of N/V, $Z(N/V, t)$, by the N-th root of $P(N, t)$:

$$\begin{aligned}
Z(N/V;\,t) = \exp\Bigg\{&-\left(1+\frac{2\sigma}{\sqrt{\pi D_r t}}\right)(4\pi D_r\sigma t)\frac{N}{V}-\left(\frac{4}{\pi}-2+2\alpha_1-2\alpha_2\right.\\
&\left.-\frac{\sqrt{3}}{\pi}\log\frac{\sigma^2}{D_r t}\right)\left(\frac{\sigma^2}{D_r t}\right)(4\pi D_r t)^2\left(\frac{N}{V}\right)^2+\left(\frac{28}{15}-\frac{16\sqrt{2}}{15}\right)\\
&\times\frac{\sigma}{\sqrt{\pi D_r t}}(4\pi D_r\sigma t)^3\left(\frac{N}{V}\right)^3+\cdots\Bigg\}.
\end{aligned} \tag{12.5}$$

This equation holds true under the condition $\sigma^2/D_r<t<V/(D_r\sigma N)$ and for this time range to exist, it must be assumed that $\sigma^3 N/V<1$.

Next, let us introduce the mean lifetime τ which stands for the average time that is required for a given particle to react in a medium with a constant density N/V, starting from the initial state of uniform distribution. From the definition of $Z(N/V, t)$, τ is calculated by the equation:

$$\tau = -\int_0^\infty t\frac{\mathrm{d}}{\mathrm{d}t}Z(N/V,t)\mathrm{d}t = \int_0^\infty Z(N/V,t)\mathrm{d}t. \tag{12.6}$$

When we consider the case where $\sigma^3 N/V<1$, the term in the exponent of Eq. 12.5, which contributes mainly to τ is $(-4\pi D_r\sigma Nt)/V$ and from

Eq. 12.6, it can be shown that τ becomes of roughly the order of $V/(4\pi D_r \sigma N)$. Substituting this τ for t in Eq. 12.5, we see that the terms due to the correlations of three and four particles have the order of $\log(\sigma^3 N/V)\cdot\sigma^3 N/V$ and $(\sigma^3 N/V)^{1/2}$, respectively. Based on this estimation, the many-body effects are regarded as small when $(\sigma^3 N)/V < 1$, that is, either σ or N/V, is small enough. In this case, the reaction rate can be approximated by the constant rate $4\pi D_r \sigma/V$; and according to the discussion in Sec. 7, the process becomes a simple stationary Markov process.

We can also obtain the survival probability in the case of r.b.c. The function $u_{k,l}$ is also given for clusters of two, three and four particles in a similar way as in the case of s.b.c. The results are summarized below:

$$u_{1,1}(0,0;t\,|\,0,0;0) = \frac{-D_r(h\sigma-1)}{2\pi^2 h}\left\{t-(1-h\sigma)\left(-\frac{1}{2D_r h^2}\right.\right.$$

$$\left.\left.+\frac{2t^{1/2}}{\sqrt{\pi D_r}\,h}+\frac{1}{Dh^2}\exp\,(Dh^2 t)\,\mathrm{erfc}\,(h\,\sqrt{D_r t}\,)\right)\right\},$$

$$u_{1,2}(0,0;t\,|\,0,0;0) = \left(\frac{D_r(h\sigma-1)}{2\pi^2 h}\right)^2\left\{1+4\frac{h\sigma-1}{h\,\sqrt{\pi D_r t}}+\cdots\right\},$$

$$u_{2,2}(0,0;t\,|\,0,0;0) = \left(\frac{D_r(h\sigma-1)}{2\pi^2 h}\right)^3\left\{-4-\left(\frac{248}{15}+\frac{64}{15}\,\sqrt{2}\right)\right.$$

$$\left.\times\frac{h\sigma-1}{h\,\sqrt{\pi D_r t}}+\cdots\right\},$$

$$u_{1,3}(0,0;t\,|\,0,0;0) = \left(\frac{D_r(h\sigma-1)}{2\pi^2 h}\right)^3\left\{-1-6\frac{h\sigma-1}{h\,\sqrt{\pi D_r t}}+\cdots\right\}. \tag{12.7}$$

Inserting these results into Eqs. 10.19 and 9.16, we obtain the survival probability for r.b.c. It is

$$P(N,t) = \exp\left[-4\pi D_r\frac{(h\sigma-1)}{h}\left\{t-(1-h\sigma)\left(-\frac{1}{2D_r h^2}+\frac{2t^{1/2}}{\sqrt{\pi D_r}\,h}\right.\right.\right.$$

$$\left.\left.\left.+\frac{1}{D_r h^2}\exp\,(D_r h^2 t)\,\mathrm{erfc}\,(h\,\sqrt{D_r t}\,)\right)\right\}\frac{NM}{V}+\frac{4}{15}(7-4\sqrt{2})\right.$$

$$\left.\times\left(\frac{4\pi D_r(h\sigma-1)t}{h}\right)^3\frac{(h\sigma-1)}{h\,\sqrt{\pi D_r t}}\frac{N^2 M^2}{V^2}+\cdots\right]. \tag{12.8}$$

The first term in the exponent is given in the exact form whose time derivative corresponds to the reaction rate of the two-body system in the case of r.b.c.   The coefficient of the second term, which is the contribution from the clusters of four particles, is given by the lowest order term in the power series of $(h\sigma-1)/(h\sqrt{\pi D_r t})$.   The terms of the third order of density are omitted here because of the fact that the contribution of clusters containing three particles is considered to be small except for the initial stage of the reaction, as discussed in the case of s.b.c.   Therefore, Eq. 12.8 holds true in the time range $(h\sigma-1)^2/(h^2 D_r)<t<hV/\{D_r|(h\sigma-1)(N+M)\}$.   From this result, it is concluded that the many-body effects become negligible when $(h\sigma-1)^3 (N+M)/(h^3 V)\ll 1$.

We have discussed many-body effects in diffusion-controlled bi-molecular reaction processes by calculating the survival probability in the form of density expansion.   The final goal of our theoretical problem, as stated in Part I, is the calculation of the probability function $U(N-n|N;\ t)$, using the results obtained.   However, it may be said that the essential feature of many-body effects has been partly revealed by the above discussion.

III.   COMPUTER SIMULATION

13.   Introduction

As we have stated in Part II, many of the works dealing with the problem of diffusion-controlled chemical reactions have treated it essentially as a two-body problem by assuming the independence of binary collisions. However, the reaction system is actually a many-body system, so there is the possibility that the two-body approximation is inadequate in cases, involving a high density and a large reaction radius of reactant molecules.

In the previous sections, a general formula of chemical kinetics from the view point of the stochastic process was presented.   A fundamental quantity, the survival probability, was obtained in the case of a diffusion-controlled bimolecular reaction process by solving the many-body problem.   Thus we have one of the keys to the analytical investigation of the many-body effect in chemical reactions.   However, it is difficult to derive the desired formula, which would directly describe the time development of the reaction system, because the probability that the number of reactant molecules is N_A' at time t (the function $U(N_A',\ t|N_A,\ 0)$ in Part II) is given by the convolution of the survival probability and probabilities of the first reaction time.   Therefore, at the present stage, our theoretical results cannot be directly compared with the experimental data and used to disclose the microscopic feature of the phenomena.

Thus, in parallel with the analytical investigation, we tried computer simulation in order to realize the stochastic properties of the diffusion-controlled chemical reactions.   We considered the irreversible bimolecular reaction process, $A+B \rightarrow C$, where A and B particles undergo Brownian's motion in a liquid medium, and a reaction occurs immediately upon collision of the two reactant particles, A and B, and the interaction between the same type particles is neglected.   In order to simulate the process of this system, we adopted the lattice model in which two kinds of particles undergo random flight on the lattice points and any pair of particles, A and B, react when they approach within the given reaction range.

14.   The model

The particles suffer random walk displacement in a three-dimensional simple cubic lattice with the volume $L \times L \times L$, where L is the number of lattice points of an edge of the cube.   We adopt the periodic boundary condition, namely, if the lattice point is specified by (l, m, n) where l, m, n takes the integer values $1, 2, \ldots, L$, point $(L+l, m, n)$ is equivalent to the point (l, m, n) and the same condition holds also for the m and n components.

Initially, $N_A^0 A$ and $N_B^0 B$ molecules are distributed on the lattice points by chance using random numbers.   This gives a sample of our system which has a uniform initial distribution.   For every unit time, each unreacted particle makes a step to its nearest neighbor site.   Each particle moves from its point (l, m, n) to one of its six nearest neighboring points $(l \pm 1, m, n)$, $(l, m \pm 1, n)$ and $(l, m, n \pm 1)$ with equal probability 1/6.   A reaction occurs when a B particle enters the " effective sphere of reaction " of any one of the A particles.   In our simulation, the effective sphere is replaced by a cube with an edge length of 2. Then, a AB pair whose coordinates are (l, m, n) and (l', m', n'), respectively, reacts when the conditions $|l-l'| \leq 1$, $|m-m'| \leq 1$ and $|n-n'| \leq 1$ are satisfied simultaneously.   A particle continues the random walk displacement until it reacts with one of the counter particles.   The movement of particles is illustrated in Fig. 1.

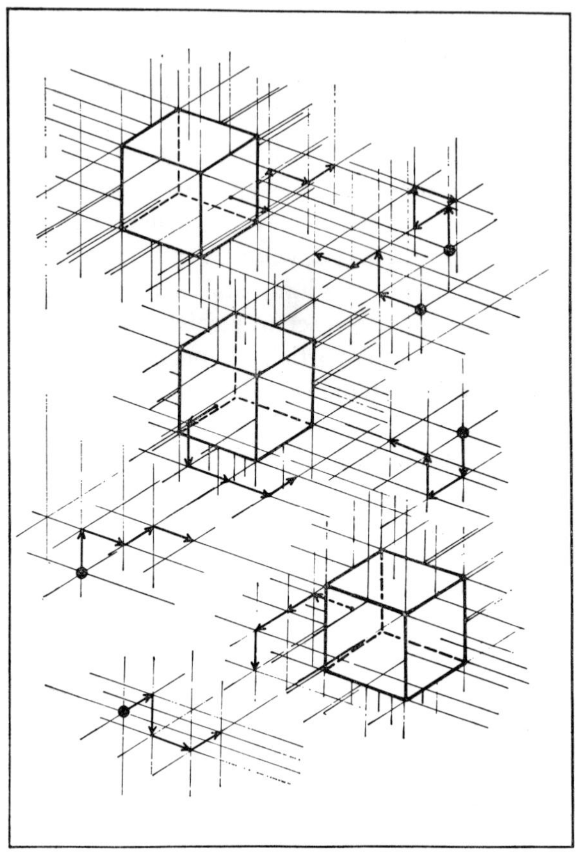

Fig. 1.   A and B particles suffer random walk displacement in the three dimensional cubic lattice.   The cubes and small solid circles represent A and B particles, respectively.   Each particle moves to its nearest neighbor position and reaction occurs when a B particle reaches the reactive surface of an A particle.

In our computer experiment, initial numbers of particles A and B were fixed at the value $N_A = N_B = 100$; and in order to see the density dependence of the phenomena, we examined several cases of different volume.   For each of these cases, the processes were traced for 50 to 100 samples.   The total number of C molecules was counted at every discrete time and the mean value of the number of C particles at time t was obtained by taking the sample average.   For example, in

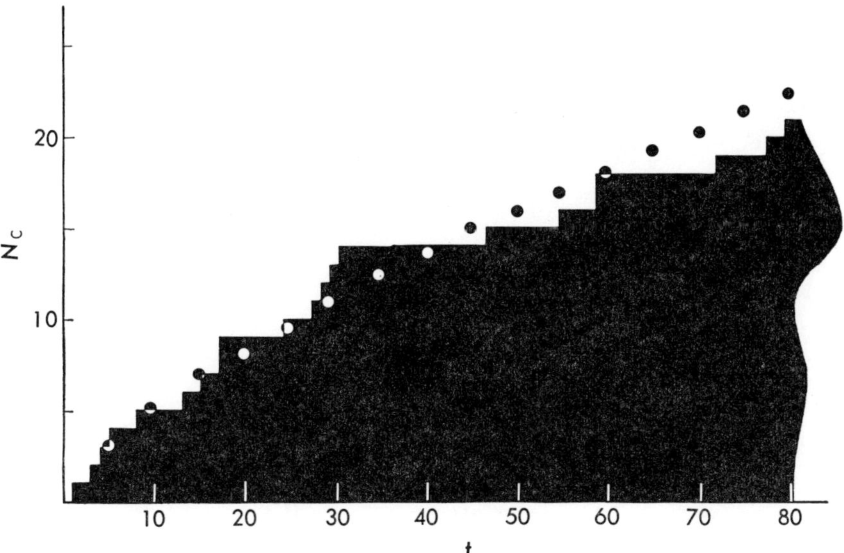

Fig. 2.   Change of the number of C particles in the case $L=60$.   Circles represent the averaged mean values of the samples.

Fig. 2 the change of the number of C particles of a sample in the early stage is shown and the mean values are also plotted by circles.

15.   Some typical results

In this section, using the numerical results of our computer simulation, we shall discuss some quantitative properties of the process of our model system of the diffusion-controlled chemical reaction.

a)   Initial number of overlapping pairs

At the initial time, $N_A{}^0=N_B{}^0=100$ particles of types A and B, respectively, are distributed on the lattice points, using the random numbers. Therefore, some pairs inevitably locate in the reaction range at the initial occasion.   If N_A A particles are placed on the lattice points at random, the number of reactive lattice points for B particles is, at most, $27 \times N_A$, because the cube of the reaction range of an A particle contains 27 lattice points.   Since the probability that a B particle falls in the reaction range of any one of the A particles is given by $27N_A / V$, then the averaged number of overlapping pairs at the initial time can be estimated as

TABLE I

L	No. of samples	Initial number of overlapping pair	
		Theoret.	Exptl.
50	50	2.16	2.03
55	50	1.62	1.64
60	100	1.25	1.34
70	50	0.79	0.74
80	50	0.50	0.60
100	50	0.27	0.18

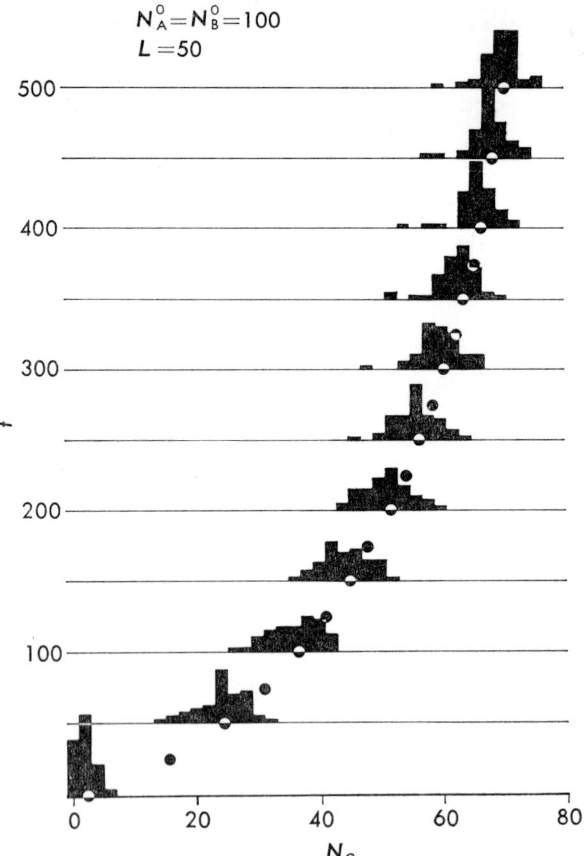

Fig. 3.   The number distribution of C particles at times $t=1$, 50, 100, . . . , 500. The circles denote the mean values of C particles at each time.

$27N_AN_B/V$. The theoretical and experimental values are given in Table I for several different values of density. It can be seen that, except in the case of $L=100$, the experimental values agree with theoretical values. This gives us one of the proofs of the uniformity of random numbers produced by the computer in this simulation.

b)   Number distribution of C molecules
The distribution function of the number of C particles as a function of time is shown in Fig. 3. The figure shows the case $N_A=N_B=100$ and $L=50$, and the time is measured by the number of steps of random walk for a particle. It can be seen that the variance or breadth of distribution increases until about the time $t=250–300$ and then gradually decreases. At an early stage of the process, the distribution has a tail at the side of the larger values of N_C; it becomes more and more symmetrical around the mean value plotted by the circle and then the tail is left on the opposite side of the mean. These qualitative properties can be compared with the theoretical distribution function derived by McQuarrie (*16, 17*) and others (*9*) from the master equation of bimolecular chemical kinetics.

c)   Survival probability and mean lifetime
In the general formula of chemical kinetics, the survival probability $P(N_A, t-t_0)$ is defined as the probability that N_AA and N_AB molecules which started the Brownian's motion at time t_0 still have not reacted at time t. This probability function can also be enumerated using the data of our simulation. Figure 4 shows the survival probabilities for several values of $N_A=N_B$. These are calculated using the data of 50 samples from the case $L=50$.

As we stated in Sec. 12, if we take only the main term, the survival probability can be expressed simply as

$$P(N_A, t) = \exp\left(-\kappa N_A N_B t\right), \tag{15.1}$$

where κ is the constant. The mean life time τ defined by

$$\tau = -\int_0^\infty t\frac{\mathrm{d}}{\mathrm{d}t}P(N_A, t)\mathrm{d}t \tag{15.2}$$

is then given simply as

$$\tau = \frac{1}{\kappa N_A N_B}. \tag{15.3}$$

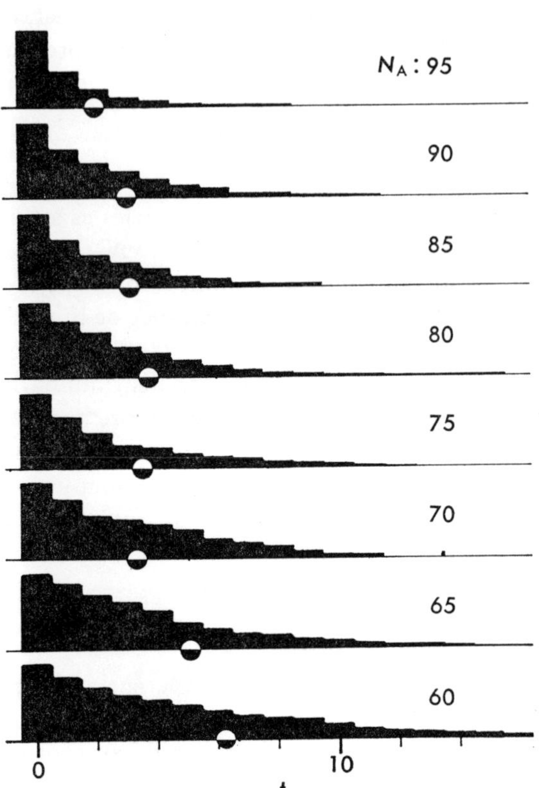

Fig. 4.   Survival probabilities and mean lifetime for several values of $N_A = N_B$.

In our computer experiment, time is discrete, so τ is expressed as

$$\tau = \sum_{n=1}^{\infty} n\{P(N_A, n) - P(N_A, n-1)\} \tag{15.4}$$

and it is plotted by circles in Fig. 4. However, unfortunately, it is impossible to draw definite conclusions from these data because the number of samples is still insufficient for the numerical analysis of many-body effect.

16. *Discussion*

The most primitive phenomenological formula of chemical kinetics is the law of mass action, which describes the time development of the irreversible bimolecular reaction process by the equation:

$$\frac{\mathrm{d}}{\mathrm{d}t}N_A = -\kappa N_A N_B \,, \tag{16.1}$$

where N_A and N_B are the numbers of A and B molecules, respectively, and κ is the rate constant. Montroll and others derived the time dependent reaction rate in a explicit form as a function of diffusion constant D_r and reaction range σ, especially, for the diffusion-controlled chemical reactions, taking into account the diffusive motion of molecules. Thus we have the equation:

$$\frac{\mathrm{d}}{\mathrm{d}t}N_A = -\frac{4\pi D_r \sigma}{V}\left(1+\frac{\sigma}{\sqrt{\pi D_r t}}\right)N_A N_B \,, \tag{16.2}$$

where D_r is the sum of the diffusion constants of two reactant molecules. In our computer simulation, $N_A = N_B$ always holds true, so the integration of Eq. 16.2 can be performed easily and we obtain

$$\frac{1}{N_A} = \frac{4\pi D_r \sigma}{V}\left(t+2\sigma\sqrt{\frac{t}{\pi D_r}}\right)+\mathrm{const}. \tag{16.3}$$

or

$$\frac{N_A^0 - N_A}{N_A^0 N_A} = \frac{4\pi D_r \sigma}{V}\left(t+2\sigma\sqrt{\frac{t}{\pi D_r}}\right), \tag{16.4}$$

where N_A^0 is the initial value of N_A.

As we discussed in Part II, any deviation from Eq. 16.4 should be regarded as a many-body effect. Now let us consider the quantity obtained by dividing Eq. 16.4 by t:

$$\frac{F(t)}{t} = \frac{4\pi D_r \sigma}{V}+\frac{8\sigma}{V}\sqrt{\frac{\pi}{D_r t}}, \tag{16.5}$$

where $F(t)=(N_A^0 - N_A)/N_A^0 N_A$. The right-hand side of this equation apparently tends to a constant value $\alpha = 4\pi D_r \sigma/V$ in the limit of $t \to \infty$. According to Eq. 16.5, the quantity:

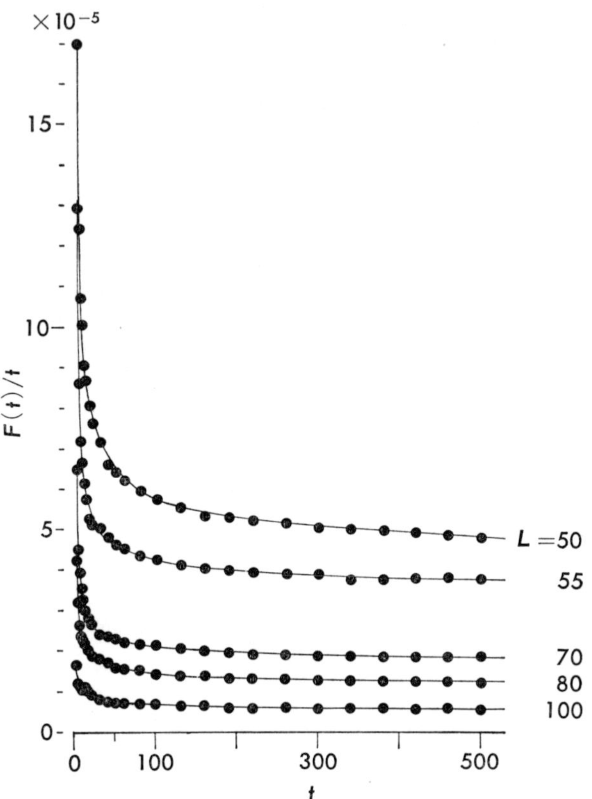

Fig. 5. $F(t)/t$ in Eq. 16.5 for several different values of volume.

$$\beta = \left(\frac{F(t)}{t} - \alpha\right)\sqrt{t} = \frac{8\sigma}{V}\sqrt{\frac{\pi}{D_r}} \tag{16.6}$$

would become independent of the time, if the many-body effect could be neglected.   Thus, the quantitative analysis of Eqs. 16.5 and 16.6 give us a way of estimating the many-body effect in our reaction system.

Figure 5 shows the values of $F(t)/t$ obtained from the data of the computer experiment for several different values of the volume.   If the reaction rate is constant, as in the case of Eq. 16.1, $F(t)/t$ must be constant.   Experimental data show the time dependence; and, as we expected, it clearly has the tendency to approach a constant value as t increases.   Table II shows these constant values which correspond to

$\alpha=4\pi D_r\sigma/V$. The experimental values of α are apparently proportional to $1/V$, and the quantity $4\pi D_r\sigma$ can be estimated at about 5.35. In the random walk model of the diffusion process, the diffusion constant is given by

$$D = n\langle r^2\rangle/6,\tag{16.7}$$

where n is the number of times of displacement per unit time and $\langle r^2\rangle$ is the mean square displacement. Though the relation expressed in Eq. 16.7 is actually plausible when $n\gg1$ and $\langle r^2\rangle\ll1$, we assume that the diffusion constant of a particle in our model has the value $1/6$, by taking $n=1$ and $\langle r^2\rangle=1$. Then the value of D_r becomes $1/3$; and using this value, we can obtain the reaction radius, $\sigma=1.28$. In our

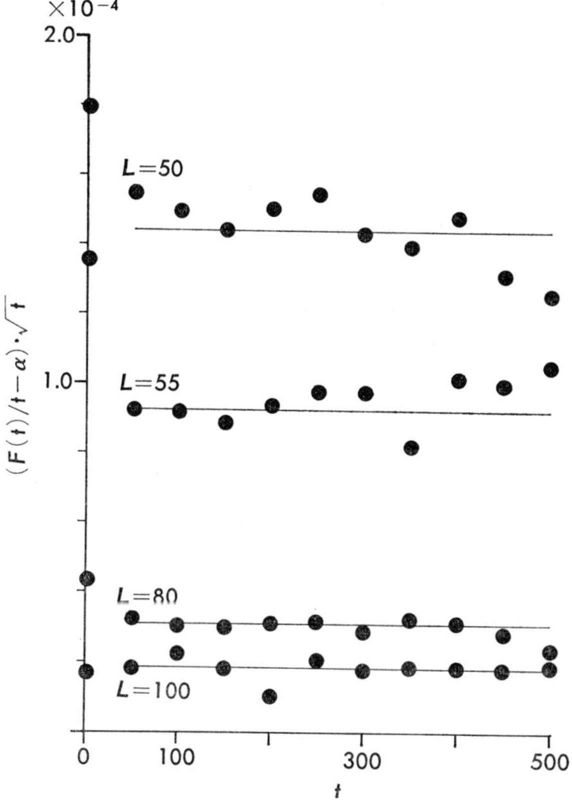

Fig. 6. $(F(t)/t-\alpha)\cdot\sqrt{t}$ in Eq. 16.6 for several different values of volume.

TABLE II

L	$\alpha\,(\times 10^{-5})$	$4\pi D_r \sigma$	$\beta\,(\times 10^{-5})$	$\beta \times V$
50	4.27	5.33	14.40	18.0
55	3.34	5.56	9.26	15.4
70	1.53	5.24	6.40	22.0
80	1.15	5.38	3.08	15.8
100	0.15	5.12	1.84	18.4

model, the effective sphere of reaction is a cube with an edge length of 2. Therefore, the value $1<\sigma<2$ seems to be quite reasonable.

The quantity given by the left-hand side of Eq. 16.6 is evaluated using the value of α given above. These values are plotted in Fig. 6. At first sight, it gives a roughly constant value for time, especially for low density cases. The roughly estimated values of β are given in Table II.

Finally, the survival probability is given in the form :

$$P(N_A, t) = \exp\{-\alpha t N_A N_B (f_0 + f_1 \alpha t (N_A + N_B) + f_2 \alpha^2 t^2 N_A N_B + \cdots)\}\,, \quad (16.8)$$

for the case that $N_A = N_B$ in Eq. 12.4 in Part II, where

$$f_0 \simeq 1 + 2\sigma/\sqrt{\pi D_r t} \qquad\qquad\qquad (16.9)$$

$$f_1 \simeq \sigma^2/D_r t \qquad\qquad\qquad (16.10)$$

$$f_2 \simeq \sigma\sqrt{D_r t}\,. \qquad\qquad\qquad (16.11)$$

In our model of the simulation, the order of α is 10^{-5}. In Eq. 16.8, the order of the first term is 1, the second term is 10^{-3} and the third term is $10^{-6}\times t^{3/2}$. Therefore, the many-body effect is really small; and Montroll's formula is plausible in the cases of those densities adopted in our computer experiments.

Further experimental computation is now going on, especially in the cases of high density, where the systematic deviation from the two-body approximation in Eq. 16.8 may be appreciably inspected.

SUMMARY

In Part I of this article, we developed a general formula to analyze the chemical reaction processes from the fundamental point of view of the stochastic process. It can, in principle, be applied to a wide range of reaction systems, starting from the description of elementary microscopic processes. For example, in Part II we discussed diffusion-controlled bimolecular reaction kinetics, using the procedure developed in Part I, and calculated the correction terms due to three- and four-body effect. In Part III, we presented the data obtained by computer simulation of the diffusion-controlled reaction kinetics, and it was shown that the many-body effect is rather small in relation to the densities of reactant molecules adopted in our computer experiment. However, at the present stage, it seems inappropriate to reach a definite conclusion about many-body effect because the number of samples used in our computer experiment was not sufficient and there was a considerable amount of sample dependent fluctuation.

The computer simulation discussed in this article was done using FACOM 230–60 in the Data Processing Centre, Kyoto University.

REFERENCES

1 R. A. Alberty and G. G. Hammes, *J. Phys. Chem.*, **62**, 154 (1958).

2 A. F. Bartholomay, *Bull. Math. Biophys.*, **20**, 97 (1958); **20**, 175 (1958); **21**, 363 (1959).

3 S. Bochner, in " Functions of Several Complex Variables," Princeton University Press, Princeton, p. 241 (1936).

4 S. Chandrasekhar, *Rev. Moder. Phys.*, **15**, 1 (1943).

5 F. C. Collins and G. E. Kimball, *Ind. Eng. Chem.*, **41**, 2251 (1949); *J. Colloid Sci.*, **4**, 425 (1949).

6 I. G. Darvey, B. W. Ninham and P. J. Staff, *J. Chem. Phys.*, **45**, 2145 (1966).

7 I. G. Darvey and P. H. Staff, *J. Chem. Phys.*, **44**, 990 (1966).

8 P. Debye, *Trans. Electrochem. Soc.*, **82**, 265 (1942).

9 K. Ishida, *Bull. Chem. Soc. Japan*, **33**, 1030 (1960); *J. Chem. Phys.*, **41**, 2472 (1964); *J. Phys. Chem.*, **70**, 70 (1966).

10 B. Kahn and G. E. Uhlenbeck, *Physica*, **5**, 399 (1938).

11 S. K. Kim, *J. Chem. Phys.*, **28**, 1057 (1958).

12 H. A. Kramers, *Physica*, **7**, 284 (1940).

13 T. D. Lee and C. N. Yang, *Phys. Rev.*, **113**, 1165 (1959).

14 J. E. Mayer, *J. Phys. Chem.*, **43**, 71 (1939).

15 J. E. Mayer and M .G. Mayer, " Statistical Mechanics," J. Wiley and Sons, New York (1940).

16 D. A. McQuarrie, C. J. Jachimowski and M .E. Russell, *J. Chem. Phys.*, **40**, 2914 (1964).

17 D. A. McQuarrie, *J. Chem. Phys.*, **38**, 433 (1963).
18 D. A. McQuarrie, *J. Appl. Prob.*, **4**., 413 (1967).
19 L. Monchick, *J. Chem. Phys.*, **24**, 381 (1956).
20 L. Monchick, J. L. Magee and A. H. Samuel, *J. Chem. Phys.*, **26**, 935 (1957).
21 E. W. Montroll, *J. Chem. Phys.*, **14**, 202 (1946).
22 E. W. Montroll and K. E. Shuler, *Advan. Chem. Phys.*, **1**, 361 (1958).
23 E. W. Montroll, " Energetics in Metallurgical Phenomena **3**," Gordon and Breach, New York, (1967).
24 R. M. Noyes, *J. Chem. Phys.*, **22**, 1349 (1954).
25 A. Rényi, *Magyar Tud. Akad. Alkalm. Mat. Int. Közl.*, **2**, 93 (1954).
26 N. Shigesada, *J. Phys. Soc. Japan*, **30**, 233 (1971).
27 K. E. Shuler and G. H. Weiss, *J. Chem. Phys.*, **38**, 505 (1963).
28 A. J. F. Siegert and E. Teramoto, *Phys. Rev.*, **110**, 1232 (1958).
29 M. Smoluchowski, *Ann. der Physik*, **48**, 1103 (1915); *Physik. Z.*, **17**, 557, 585 (1916); *Z. Physik. Chem.*, **92**, 129 (1917).
30 P. J. Staff, *J. Chem. Phys.*, **46**, 2209 (1967).
31 T. R. Waite, *Phys. Rev.*, **107**, 463 (1957); *J. Chem. Phys.*, **28**, 103 (1958); **32**, 21 (1960).
32 J. H. Wang, *J. Phys. Chem.*, **59**, 1115 (1955).
33 G. H. Weiss, *Advan. Chem. Phys.*, **13**, 1 (1967).
34 J. Yguerabide, M. A. Dillon and M. Burton, *J. Chem. Phys.*, **40**, 3040 (1964).

Received for publication, August 27, 1970.

CONTRIBUTOR SKETCHES

The Lateral-Line Organ of Shark as a Chemoreceptor
Yasuji KATSUKI is a professor of physiology at Tokyo Medical and Dental University. He graduated from University of Tokyo Medical School in 1931. From 1934 to 1949 he was a research assistant and then a lecturer in the Department of Physiology, University of Tokyo. He spent several years working at the Central Institute for the Deaf in St. Louis, the Massachusetts Institute of Technology, the University of California, and the University of Hawaii. Professor Katsuki is an authority on sensory physiology, especially auditory mechanisms. He was awarded the Asahi Newspaper Prize of Cultural Merits in 1962 and the Japan Academy of Science Prize in 1963.

Toru HASHIMOTO is a research associate at Tokyo Medical and Dental University. He received the M.S. degree in electrical engineering from Kyoto University in 1961 and has been affiliated with Professor Y. Katsuki 's laboratory since 1968.

Keiji YANAGISAWA is a research associate at Tokyo Medical and Dental University. He has been associated with Professor Y. Katsuki in investigating the auditory mechanism in insects and cats.

Ribonuclease T_1—Structure and Function
Fujio EGAMI is a professor of biochemistry at University of Tokyo. Following his graduation from the Department of Chemistry, University of Tokyo in 1933, he spent two years in Strassbourg and Paris. From 1943 to 1960 he was a professor of chemistry at Nagoya University. Professor Egami's wide range of contributions to biochemistry includes sulfatases, nitrogen metabolism, bacterial toxins, ribonucleases, and glycosidases. In 1967 he was awarded the Asahi Newspaper Prize of Cultural Merits. Professor Egami has been President of the Japan Science Council since 1969 and has served as a council member of the International Union of Biochemistry since 1961.

Tsuneko Uchida is a research associate in the Department of Biochemistry and Biophysics, University of Tokyo. She graduated from the Department of Chemistry, Nagoya University in 1954, and since then she has been closely associated with Professor F. Egami. Her main interest is the biochemistry of proteins and nucleic acids.

Kenji Takahashi is a research associate in the Department of Biochemistry and Biophysics, University of Tokyo. In 1965 he worked with Drs. S. Moore and W. H. Stein at Rockefeller University. In the same year Dr. Takahashi was awarded the Prize of the Biochemical Society of Japan for his study on the amino acid sequence of ribonuclease T_1. His main area of study is the primary structure and function of enzymes.

Polymerization of Flagellin and Polymorphism of Flagella
Sho Asakura is a professor of molecular biology at Nagoya University. He graduated from the Department of Physics, Nagoya University in 1952. Dr. Asakura's achievements include work on ATPase action of F-actin under sonic vibration and the study of *in vitro* reconstruction of flagella, the results of which he presented as an invited speaker at the 3rd International Congress of Biophysics in 1969.

Spin Changes in Hemoproteins
Takashi Yonetani is a professor of physical biochemistry at the Johnson Research Foundation, University of Pennsylvania. He is a graduate of Osaka University. Influenced by Professor K. Okunuki, he has been deeply involved in the physicochemical study of cytochromes, especially cytochrome oxidase. He worked with Professor B. Chance in 1962 and with Professor H. Theorell in 1966. Professor Yonetani is a recipient of the Career Development Award from the U.S. Public Health Service.

Tetsutaro Iizuka is a visiting assistant professor of biophysics at the Johnson Research Foundation, University of Pennsylvania. He graduated from the Department of Physics, University of Tokyo in 1964 and after spending three years at Osaka University, he joined Professor T. Yonetani's group to conduct biophysical research on cytochrome C peroxidase.

VOLUME 2

Phase Transition in Membrane with Reference to Nerve Excitation
Akira WATANABE is a professor of physiology at Tokyo Medical and Dental University. He has published a number of articles on the physiology of excitation in the Proceedings of the National Academy of Sciences, U. S. and other journals.

Ichiji TASAKI is the chief of the Neurobiology Laboratory at the National Institute of Mental Health, Bethesda. He was educated at Keio University where he was introduced to neurophysiology by Professor G. Kato. From 1942 to 1951 he was a professor of physiology at Keio Medical College and since 1951 he has been in the United States. Dr. Tasaki is a well-known neurophysiologist and has published numerous articles. He was awarded the Superior Service Award of the Department of Health, Education, and Welfare in 1968 for his work on the elucidation of the basic mechanism of nerve excitation.

Yonosuke KOBATAKE is an associate professor at Osaka University. He graduated from the Tokyo Institute of Technology in 1950 and has been engaged in physicochemical studies on biological membranes. Dr. Kobatake received the Society Award of the Japanese Society of High Polymer Science in 1968.

One-Electron and Two-Electron Transfer Mechanisms in Enzymic Oxidation-Reduction Reactions
Isao YAMAZAKI is a professor of biophysics at the Institute for Applied Electronics, Hokkaido University. He graduated from the Department of Chemistry, Hokkaido University in 1949. His doctorate thesis was entitled "The oxidoreductive feature of intermediate formed in the reaction of peroxidase." Professor Yamazaki's main interest is the mechanism of enzymic electron transfer reaction.

The Electrogenic Sodium Pump
Kyozo KOKETSU is a professor of physiology at Kurume University Medical School. He graduated from Kyushu University in 1946 and has worked on a research project centered around the physiology of cell membranes.

Structure of Tropomyosin and Its Crystal
Tatso Ooi is a professor at the Institute for Chemistry, Kyoto University. Since graduating from the Department of Physics, Nagoya University in 1947, Professor Ooi has worked on the physical chemistry of proteins, especially actin and tropomyosin. He has worked twice with Dr. H. A. Scheraga at Cornell University.

Sugie Fujime is a research associate in molecular biology at Nagoya University. She graduated from the Department of Physics, Kobe University in 1962. Her main interest is protein crystal and its phase transition.

Stochastic Theory of Reaction Kinetics
Ei Teramoto is a professor of mathematical biophysics in the Department of Biophysics, Kyoto University. He graduated from the Department of Physics, Kyoto University in 1947. Professor Teramoto is known for his study on statistical mechanics, especially lattice dynamics of high polymers. Recently, he has become interested in the mathematical analyses of more biological aspects, *e.g.*, population and behavior biology.

Nanako Shigesada is a research associate in the Department of Biophysics, Kyoto University. She obtained her Ph. D. in 1969 and her thesis was concerned with the statistical mechanics of chemical reaction.

Hisao Nakajima is a graduate student in the Department of Physics, Kyoto University where he is studying mathematical biophysics under the supervision of Professor E. Teramoto.

Kouji Sato is a graduate student of the Department of Physics, Kyoto University. His main interest is automata theory.

PUBLISHED PAPERS

VOLUME 1

Yasuji KATSUKI, Toru HASHIMOTO and Keiji YANAGISAWA, The Lateral-Line Organ of Shark as a Chemoreceptor

Kenji TAKAHASHI, Tsuneko UCHIDA and Fujio EGAMI, Ribonuclease T_1—Structure and Function

Sho ASAKURA, Polymerization of Flagellin and Polymorphism of Flagella

Tetsutaro IIZUKA and Takashi YONETANI, Spin Changes in Hemoproteins

FORTHCOMING PAPERS

Satoru FUJIME, Quasi-Elastic Scattering of Laser Light—A New Tool for the Study of Biological Macromolecules

Hiroshi KOBAYASHI, Metal Enzyme Model

Takamitsu SEKINE, Myosin ATPase

Akira TAKEUCHI, Actions of Transmitters on Synaptic Membranes

Mitsuo NISHIMURA, Energy Transfer in Photosynthetic Membrane Systems

Hiromichi MORITA, Primary Process of Chemoreception

Setsuji HATANO, Contractile Proteins from the Myxomycete Plasmodium

Haruo OZEKI, Mutant tRNA

Susumu HAGIWARA, Ca Spike

Shoten OKA, Theoretical Studies on the Rheology of Blood and Blood Vessels

Uichiro KISHIMOTO, Characteristics of Excitable Membranes

Chikayoshi NAGATA, The Electronic Structures of the Bases of Nucleic

Acids in Relation to Biological Activity

Taro HURUKAWA, An Electrophysiological Analysis of the Activities of Hair Cells in Goldfish's Inner Ear

Tateo YAMANAKA, The Evolution of the Cytochrome c Molecule

Reiji OKAZAKI, Mechanism of DNA Replication

Shigeo YOMOSA, Electronic Processes in Polypeptides and Proteins

Eiichi FUKADA, Piezoelectricity in Biological Substances

Akiyoshi WADA, Internal Order and Rigidity of Polypeptides and Polynucleotides

Hiroshi WATARI, The Electronic State of Non-heme Iron Proteins

Shiro AKABORI, Origin of Life